MYSTERIES
OF THE HUMAN BODY

>>

THIS IS A CARLTON BOOK

Design copyright © 2005 Carlton Publishing
Group
Text copyright © 2005 Carlton Publishing
Group

This edition published in 2005 by Carlton
Books Ltd
A Division of the Carlton Publishing Group
20 Mortimer Street
London
W1T 3JW

A CIP catalogue for this book is available
from the British Library.

ISBN 1 84442 363 8

Executive Editor: Amie McKee
Senior Art Editor: Zoë Dissell
Design: Anita Ruddell
Picture research: Steve Behan
Production: Lisa Moore

Printed in Dubai

MYSTERIES
OF THE HUMAN BODY

GORDON THOMAS

CARLTON
BOOKS

CONTENTS

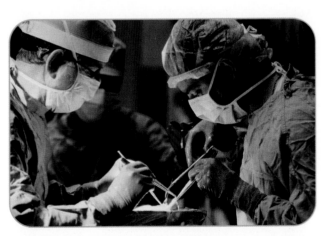

INTRODUCTION

For all of the microchip-inspired technical innovations that have flooded modern society, no piece of machinery is more complex than the human body – a vast network of 13,000 nerve cells, 650 muscles, 206 bones, 100 joints and 60,000 miles of blood vessels. Yet although we have made tremendous strides in the various fields of medicine in recent years, there is still much about the body that we do not understand. Take the brain, for example. We know which areas are responsible for hearing, language, movement and sensation, but the brain's elaborate circuitry holds the answers to many more still unsolved riddles, such as why do we sleepwalk and why are yawns contagious? This is part of the ongoing fascination with the human body – medical science can carry out lifesaving transplants yet remain baffled by everyday subjects.

Thanks to the advances in surgery and patient care, more medical stories than ever have a happy ending. We have almost become accustomed to miracle cures yet some tales featured within these pages still seem to defy logic: the healthy baby that grew in the mother's liver; the man who regained consciousness after 19 years in a coma; the boy who was pregnant with his twin; the man whose head was reattached after a car crash; the girl who gave birth at the age of five; and the woman who survived after all her skin fell off. Then there are seemingly incomprehensible phenomena, such as the Russian girl with X-ray vision, electric people who claim to be able to knock out streetlights at a glance, and the curious Foreign Accent Syndrome. And of course there are the questions that have stumped the finest scientific minds for centuries. So why is it that some arthritis sufferers can predict rain through their joints? What exactly are eyebrows for? And why can't we tickle ourselves?

As well as examining unusual allergies and charting current breakthroughs in medicine, I have endeavoured to unravel the aforementioned mysteries. In some instances, new studies have brought about possible explanations, but others remain as elusive as ever. In truth, no matter how much time and expertise we devote to the subject, we may never solve every mystery of the human body. For it seems that the more we learn about our inner workings, the more unanswered questions the body throws up.

1

MIRACLE BABIES

Until around 50 years ago giving birth was a dangerous procedure, for mother baby alike. Women of all social classes commonly died in the course of childbirth. While maternal mortality still remains worryingly high in developing countries, elsewhere medical advances in monitoring and treating the various complications of pregnancy and childbirth have resulted in a marked decline in the mortality rate among women in developed countries. In 1950 maternal mortality in the UK was around 60 per 100,000 pregnancies; now it is down to less than 10 per 100,000. Although a vast majority of births pass without alarm, there remain a number of potential complications caused by such conditions as high blood pressure, anaemia, or ectopic pregnancies where the foetus develops outside the uterus. In bygone centuries these conditions would often have proved fatal, but these days many are now treatable and babies who would once have perished are now born perfectly healthy. Every now and then, however, a truly remarkable case still comes along that confounds even modern medical science by producing a healthy baby against all the odds. Mothers and fathers who had abandoned any hope of a successful birth suddenly find themselves cradling a tiny bundle of joy. To relieved parents, these babies are nothing short of miracles.

The Baby That Survived a Termination

Suffering from a severe form of the potentially fatal condition pre-eclampsia, Norelle Smith was told by doctors that she would die if she did not abort her baby at 26 weeks. Reluctantly, she agreed to terminate the pregnancy and drugs were used to induce the baby, who was not expected to live longer than a few hours. Defying all the odds, baby Natasha survived and, despite growing for just 22 weeks in her mother's womb and weighing only 1lb 4oz (567g) at birth, was able to leave hospital four months later, a small but healthy child. Cradling Natasha in her arms at home in Oban, Scotland, a delighted Norelle said: "She is gorgeous and a medical miracle".

For Norelle and her partner Sandy Cameron it was an unexpected and happy ending to a traumatic eight months. Norelle had pre-eclampsia, which can cause liver and kidney problems for a pregnant woman, while expecting their first child, Sean, who was born five weeks early. Pre-eclampsia is a serious condition in which the woman suffers hypertension (high blood pressure), oedema (an accumulation of fluid in tissues) and proteinuria (protein in the urine) during the second half of the pregnancy. It affects about seven per cent of pregnancies, but is more common in first pregnancies. After the problems with her first baby, Norelle was assured that it could not happen again, but when she fell pregnant with Natasha she had a threatened miscarriage on Boxing Day 2003. Then in February 2004 she was told there was an 85 per cent chance that Natasha had spina bifida and was advised to terminate the pregnancy, she decided to continue. "The pre-eclampsia was discovered on April 4," said Norelle, "but checks showed I'd had it from day one. The threatened miscarriage and spina bifida scare were all down to pre-eclampsia."

The toxic pre-eclampsia had resulted in Natasha being starved of oxygen and nutrients in the womb, it was also putting Norelle's own life at risk. On April 26 doctors at Glasgow's Queen Mother's Hospital told Norelle that the baby would either be born dead or would live for just a few hours. As Norelle's health deteriorated, she was taken to the high-dependency unit.

"I had the option to deliver Natasha myself, but was told she wouldn't survive. The alternative was to have a Caesarian section. This would have destroyed my uterus, my chance of future children and would probably still have killed my baby. After a test on April 27, the decision was taken out of my hands. My liver and kidneys were about to pop and the doctors said I'd have to end the pregnancy. My life was on the line. I was devastated because the baby was planned and desperately wanted."

Norelle was induced that day, but when nothing happened she was given stronger tablets. Although she had repeatedly been told that there was no chance of the baby surviving, Norelle refused to accept this because deep down she could feel it moving. "From 22 weeks, every scan showed her getting smaller," said Norelle, "but I refused to believe I would lose her."

Norelle was alone with Sandy and her mother when she felt contractions. Her mother immediately sought out the midwife who arrived just as Natasha fell out on to the table in an intact membrane. That "balloon" of life also contained the afterbirth and placenta and almost certainly saved Natasha's life, as did the quick thinking midwife who saw that Natasha

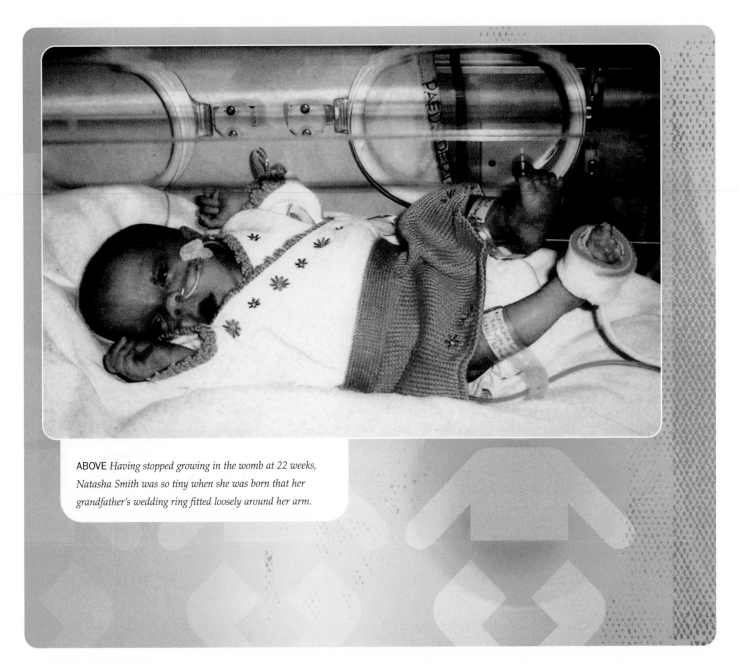

ABOVE *Having stopped growing in the womb at 22 weeks, Natasha Smith was so tiny when she was born that her grandfather's wedding ring fitted loosely around her arm.*

was alive and rushed the membrane containing Natasha away to intensive care. There the paediatric team removed the baby from the membrane and ventilated her.

Natasha measured just 6in (15.2cm) long and was so frail that her grandfather's wedding ring fitted loosely around her arm. She had stopped growing at 22 weeks and had been forced from the womb 14 weeks early. When she was brought to her mother's bedside in an incubator, doctors again warned that she was not expected to live long.

Norelle recalled: "All I could see in the incubator were these two big, beautiful eyes. We were told it would just be a couple of hours and they gave me photos of her. But then the hours turned to days and the days turned to weeks. We were told she'd be badly deformed and brain-damaged because she was starved of oxygen but there is nothing wrong with her. She came out so perfect. She was small, but she was a formed baby. She had eyelashes, nails, hair on her head. She opened her eyes and she could cry. She was breathing by herself."

By the age of eight months, Natasha was tipping the scales at a healthy 8lb 13oz (3.889kg). Although she still wore tiny baby clothes, doctors said that she was developing perfectly. She was eating solids and

ABOVE *Proud mum Norelle Smith with baby Natasha, whose frailty at birth meant that doctors held out little hope for her survival.*

was doing considerably more than expected for a baby of her size and age. "She's amazing," said Norelle. "She's always laughing and smiling. She loves the sound of her own voice. When she went down to Glasgow for routine tests, they blew bubbles that she not only saw, but caught and burst. Rattles were used to test Natasha's hearing, but we've always known there's nothing wrong there."

Consultant Dr Janice Gibson echoed those views: "Natasha was a wonderful surprise. She was so small we didn't expect her to have any chance of survival but the stress she suffered inside the womb meant she had already developed coping measures before being delivered. She's a true miracle."

Naturally, the survival of a baby that had stopped growing at 22 weeks reignited the abortion debate – the current time limit for UK terminations being 24 weeks. As renewed calls were made for the time limit to be lowered, Norelle said: "I don't agree with abortions really. Medically, I suppose people have reasons, but it's a baby from day one. Seeing Natasha made us realise how wrong it is to choose to terminate a baby as late on as that."

Sandy Cameron added: "We were told neither of them would survive or at least one of them wouldn't survive and that was going to be Natasha. It's hard to take in, when you've got both of them here, that you were faced with having neither of them."

The Baby That Grew in Her Mother's Liver

Twenty-year-old Ncise Cwayita had already given birth to one child normally and there was no reason to suspect that her second pregnancy would be anything out of the ordinary. Some months into her pregnancy she had been checked at a clinic in Cape Town, South Africa, and told that everything was fine. But then, in May 2003, she was referred to Somerset Hospital with high blood pressure. When a scan was done, startled staff saw that Ncise's womb was empty, even though she was just a week off full term. Instead, in a case that made newspaper headlines around the world, the baby had been developing in her mother's liver.

On duty in the labour ward at Somerset Hospital that night was sixth-year medical student Lindsey Bick. "She looked like a normal pregnant woman," recalled Lindsey. However, on examining her, she couldn't find the baby's head, usually easy to detect in the pelvis at 39 weeks. Also, the baby's bottom was very high in the abdomen. Mystified, she alerted her supervisor who conducted an ultrasound scan, which confirmed Lindsey's misgivings. There was no sign of the baby's head. Moreover, the uterus was empty.

Ncise was transferred to Groote Schuur Hospital where subsequent scans confirmed an advanced extra-uterine pregnancy and the location of the placenta in the upper abdomen. "We see about five or six cases of extra-uterine pregnancy a year," said senior obstetrician Dr Bruce Howard, a specialist in gynaecological cancers and difficult surgery. As it was such a rare occurrence, he remembers looking forward to the operation, but this one had another twist to it.

Inside the operating theatre, the drama deepened when the Caesarean section revealed a further puzzle: all Ncise's organs were in place (even the empty uterus), but there was no baby. As news spread through the hospital, some 30 doctors and medical students gathered around the patient to witness what was about to unfold. Instead of a foetus, Dr Howard found a "massively enlarged" liver and placenta. "It was very frightening," he said. "A time bomb waiting to explode."

When an egg is fertilised, it normally travels down the fallopian tube to the womb where it implants and grows, but sometimes the embryo implants in the fallopian tube – a standard ectopic pregnancy. In around one in 100,000 pregnancies, the embryo falls out of the fallopian tube and can implant itself anywhere in the abdomen. In extremely rare cases, such as this one, the embryo attaches itself to the surface of the liver, a very rich source of blood, and then grows into the liver, pushing the mother's liver cells aside. Although the baby is protected because it is within the placenta, it does not have the usual protection of the muscular wall of the womb and is at greater risk in the abdominal cavity. The chances of survival are slim – most babies in extra-uterine pregnancies die within a few weeks. Abdominal pregnancies also pose a serious risk to the mother.

Having discovered that Ncise's baby was in the liver, Dr Howard called in liver surgeon Professor Jake Krige. The liver, about the size of a rugby ball, is a highly dangerous organ on which to operate because it is rich in blood vessels and bleeds easily. The placenta, in its amniotic sac, lined Ncise's liver. To remove it would have resulted in catastrophic bleeding, so Professor Krige had to resort to improvisation in order to find a way in to deliver the baby. By chance, at the base of the organ he and Dr Howard found a "window" 1.97in (5cm) diameter, a small area where the placenta and amniotic sac weren't attached. It offered the sole point of entry. An incision was made and the baby's delivery began, feet first. "It was extraordinary," said Professor Krige. "A breech delivery from the liver."

First the baby's left foot came out, then her right, then her body, followed by her arms and finally her head. But there was still much work to be done. The baby was distressed and needed resuscitation, and the placenta had started to bleed. Fortunately the specialists were able to stem the flow. There then remained the question of what do with the placenta and the amniotic sac. It was decided to leave them attached to the liver because to remove them would have posed too great a threat to the mother's life and in a month or two they would be absorbed back into the body anyway.

Baby Nhlahla (the Zulu word for "luck") weighed in at a very healthy 6lb 3oz (3.91kg) and, although having been put on oxygen immediately after the birth, was able to breathe without aid two days later. There have only been 14 previous instances of a baby developing inside the mother's liver instead of in the womb and, because of bleeding complications, Nhlahla was only the fourth to survive such a pregnancy. As Professor Krige summed up: "She is truly a miracle baby".

All the more so because, as Dr Howard pointed out: "If Ncise had had access to ultrasound from the beginning, as is the case in first-world countries, the pregnancy would have been terminated. Now we have a healthy mother and baby."

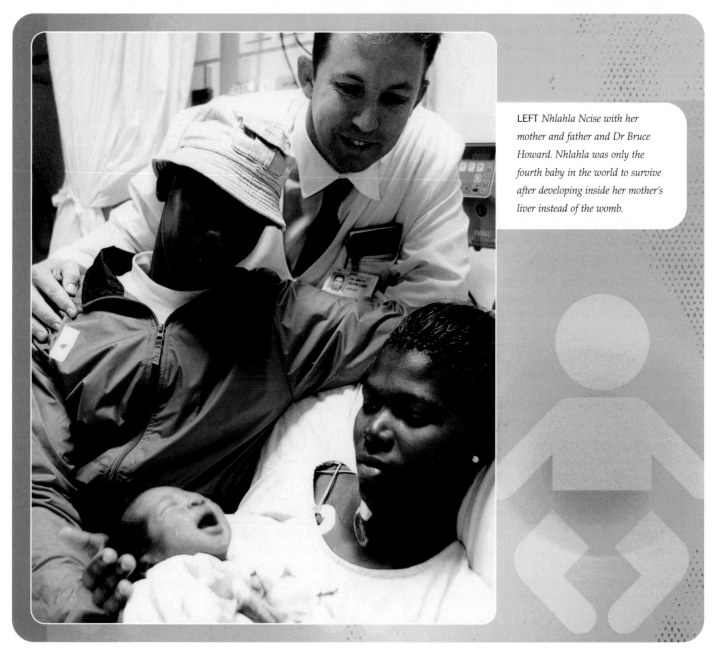

LEFT *Nhlahla Ncise with her mother and father and Dr Bruce Howard. Nhlahla was only the fourth baby in the world to survive after developing inside her mother's liver instead of the womb.*

The Woman Who Woke From a Coma to Find She Had Given Birth

The last thing Joanne Roberts remembered before she slipped into unconsciousness was being unable to see. But when she woke from a coma, she found that she was the mother of a baby boy – and she hadn't even known that she was pregnant. What makes Joanne's story all the more remarkable is that she had been told she would be unable to conceive after being injured in a car crash as a teenager.

Joanne had been feeling unwell for some time. She complained of a sore throat and thought she had tonsillitis, but then she suffered a fit and collapsed at her home in St Helens, Merseyside. She was rushed to Whiston Hospital where she was given painkillers and put into a coma to prevent further seizures. While she lay unconscious, doctors discovered that she was pregnant and had the potentially life threatening disorder, pre-eclampsia. She was rushed to theatre and underwent an emergency Caesarean section. Luke was born without a heartbeat and was medically dead for several minutes before doctors revived him. Born two months premature, he weighed just 4lb 2oz (1.871kg) and had to spend time in intensive care.

Joanne woke up some days later and was puzzled when staff put a tiny baby into her arms.

She said later:

"I thought someone was pulling my leg and dumping their baby on me. Until I was out of the coma I had no idea I was pregnant or that I'd had a baby. I just did not have any of the symptoms, apart from swollen ankles. The thought of having children just never entered my mind because I was told I would not be able to have any, so I never dreamed pregnancy could have been a possibility."

Luke was born in July 2004 – the day before Joanne's birthday – and by September the family were able to reflect on the surprise birth.

Joanne's mother said:

"We were warned that she could die. It was so frightening. Doctors told us to wait in this small room after she was rushed into hospital. They worked on her and I really did think it was going to be bad news. At one stage the nurse came into us and told us Joanne was fitting and they could not bring her out of it. We were told she could have a stroke or even die. I just prayed for Joanne's life, then when she woke up in intensive care and said her dad would kill her, I knew she had no brain damage and would be OK. Both Joanne and Luke have been very, very lucky. We are just so grateful to the hospital for everything they did."

Joanne, who calls Luke her "little miracle", was equally grateful to friends and family who rallied round and offered support. "Obviously we didn't have a single item of baby clothing in the house, but people have bought clothes for Luke and lots of other essential equipment, which I would never have thought of. After all, I did not have nine months to prepare for my arrival like other mums."

A Mother Aged Five

Lina Medina's parents thought their five-year-old daughter was suffering from a massive abdominal tumour. When the shamans in their remote village of Ticrapo high up in the Peruvian Andes could find no cure – even ruling out a local superstition where it was believed that a snake grew inside a person until it killed them – her father carried her to hospital in the nearby town of Pisco. There, doctors broke the astonishing news that Lina's abdominal swelling was due to the fact that she was pregnant. She was transferred to a hospital in Lima where, just over a month later on May 14, 1939 (coincidentally the date on which Mother's Day was celebrated that year), her child was born by Caesarean section. Aged five years, seven months and 21 days, Lina made medical history by becoming the youngest known mother in the world – a dubious distinction that she still holds to this day.

Her son, Gerardo, weighed 5lb 14oz (2.665kg) at birth and was named after Dr Gérado Lozada, one of the doctors who carried out the Caesarian section. The baby was remarkably healthy and a few days later, mother and child were able to leave hospital. Lina's father was jailed briefly on suspicion of incest, but he was released due to a lack of evidence. The authorities were never able to determine who was the real father of Lina's child since the young mother could not give precise responses. The boy himself was raised believing that Lina was his sister, but when he was ten, after taunting from schoolmates, he discovered that she was in fact his mother.

Dr Lozada, chief physician at the Hospital of Pisco, first came into contact with Lina in early April 1939 when she was brought to him for evaluation of the supposed tumour. Her medical history revealed that she had been having regular periods since the age of two and a half. By the time she was four years old she had already developed breasts as well as pubic hair. Hers was an extreme case of precocious puberty, where physical and hormonal signs of pubertal development appear at an earlier age than is thought normal. Puberty usually occurs in girls between the ages of eight and 13 and in boys between the ages of nine and 14, but some research indicates that puberty may begin as early as seven years in Caucasian girls and six years in black girls. In general, however, any girl that develops breasts, armpit or pubic hair, or experiences menstruation before the age of eight is considered to be demonstrating precocious puberty. The cause of the condition is not fully known. It is, however, thought to be genetic and, especially in girls, may also be partly attributable to increased body fat.

When Dr Lozada carried out a further examination on Lina, he detected a foetal heartbeat, and an X-ray confirmed that she was pregnant. A biopsy performed on one of Lina's ovaries from a sample removed at the time of the Caesarean section revealed that she had the ovaries of a fully mature woman. Edmundo Escomel, one of Peru's leading physicians of the day, suggested that the reason for the girl's early fertility did not lie in the ovaries themselves, but must have stemmed from an extraordinary hormonal disorder of pituitary origin.

The head of the hospital where Lina's baby was delivered described the case as "truly astounding". As the story reached the American press, a Chicago doctor recalled the case of a Russian girl who had become a mother at the age of six and a half and apparently had the physical development of a girl of ten or 12 years old. There was great excitement in Lina's story, which was verified by such bodies as the American College of Obstetricians and Gynaecologists, and the family were offered $1,000 for Lina and the baby to be exhibited at a world fair in New York. With eight other children to raise in Peru's poorest province the money might have proved tempting, but before any deal could even be considered, the Peruvian government stepped in and announced that Lina and her baby were in "moral danger". The government promised financial aid, but none ever materialized and she was condemned to a life of poverty. According to obstetrician Jose Sandoval, who has studied the case, Lina was a psychologically normal child who displayed no other unusual medical symptoms. Perhaps not surprisingly, she preferred playing with dolls rather than her own child.

Lina married in 1972 and had a second son, 33 years after her first. Then in 1979 Gerardo died at the age of 40 from a disease that attacks the body's bone marrow, although there was no obvious link between his illness and the fact that his mother had been so young at his birth. Lina and her husband currently live in a crime-ridden district of Lima called "Little Mexico", where she maintains a low profile, refusing all requests to discuss the past. Sandoval's book has raised fresh interest in this remarkable case and has spurred the Peruvian government into belated action. However, Lina's husband, Raul Jurado, said his wife remained sceptical. "She got no help in 1939 that I know about," he said recently. "She thinks government never deliver."

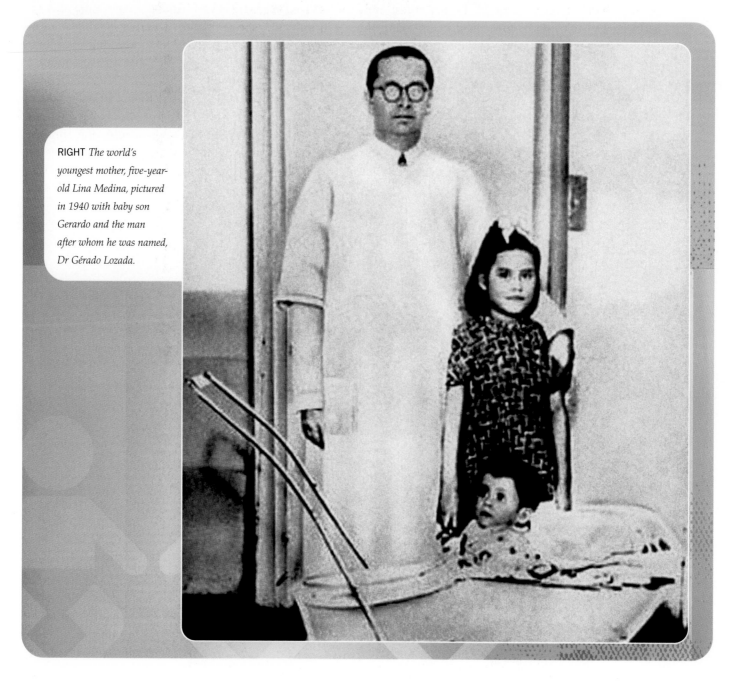

RIGHT *The world's youngest mother, five-year-old Lina Medina, pictured in 1940 with baby son Gerardo and the man after whom he was named, Dr Gérado Lozada.*

The McCaughey Septuplets

Born nine weeks prematurely in Des Moines, Iowa, USA on November 19, 1997, the McCaugheys are the world's first surviving set of septuplets. Conceived as a result of fertility drugs, the seven babies were delivered by Caesarean section in the space of six minutes and astonished doctors with their health. "All the babies are so well grown, so well developed," reported Dr Paula Mahone shortly after the multiple births. "I would consider this a miracle."

Septuplets do not occur naturally. Their mother, Bobbi McCaughey, of Carlisle, Iowa, had been taking the fertility drug Pergonal, which was prescribed because she and her husband Kenny had trouble conceiving their first child Mikayla, who was born in 1995. When first told that she was carrying seven foetuses, Mrs McCaughey admitted to a feeling of "sheer terror". Doctors suggested the possibility of "reduction" – aborting some foetuses to allow others room to grow – but the McCaugheys, who met at a Bible college, refused to consider the option. "Any child is a gift from God," she said later, "whether it's one at a time or seven at a time. It didn't take very long – just a few weeks – to get used to the idea that we were going to have a very big family."

The couple said they used their Christian faith to help them through difficult times during the pregnancy. One week, doctors could only find six heartbeats; on another occasion one foetus was not bathed in enough amniotic fluid, the "waters" that

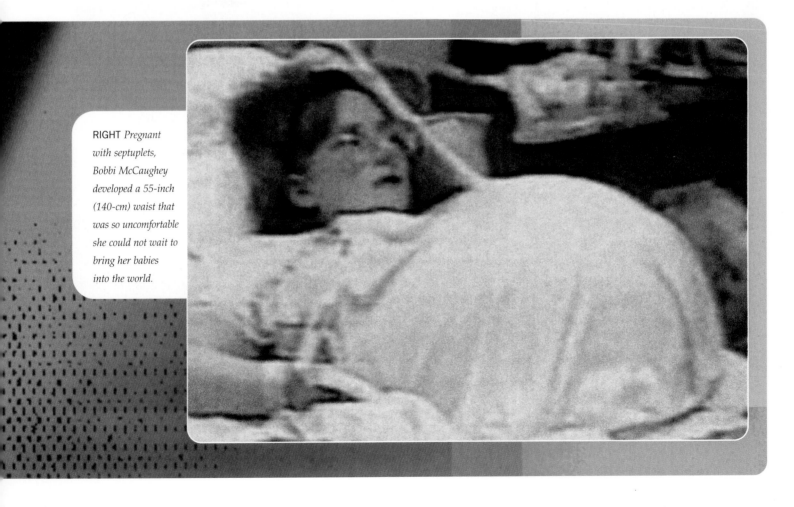

RIGHT *Pregnant with septuplets, Bobbi McCaughey developed a 55-inch (140-cm) waist that was so uncomfortable she could not wait to bring her babies into the world.*

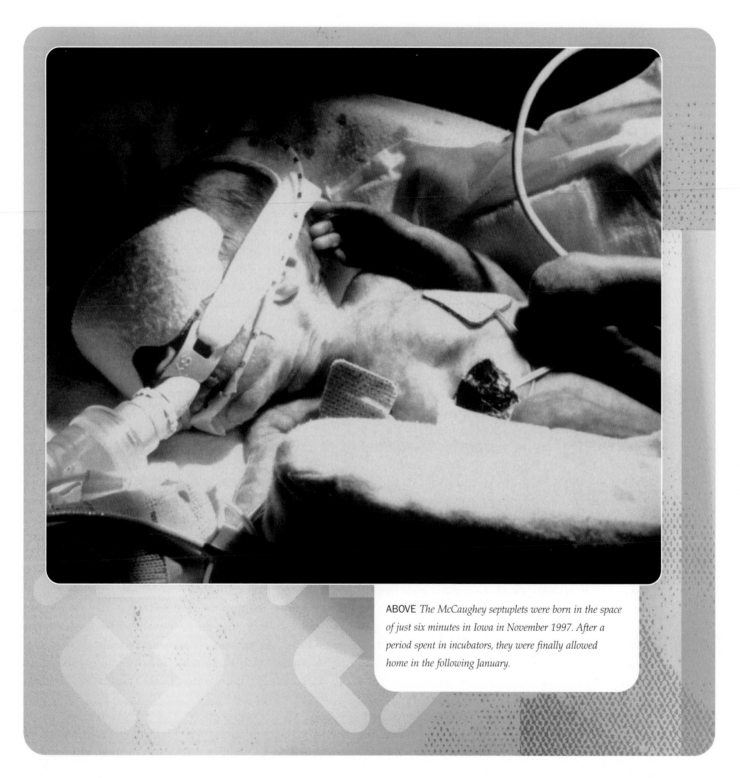

ABOVE *The McCaughey septuplets were born in the space of just six minutes in Iowa in November 1997. After a period spent in incubators, they were finally allowed home in the following January.*

cushion the foetus against pressure from internal organs and protect it from any injury caused by the mother's movements. Happily, in both cases, the danger passed.

In October 1997 Bobbi McCaughey was admitted to the Iowa Methodist Medical Centre so that doctors could constantly monitor the condition of the babies and ensure that she was receiving sufficient rest and nutrition. The ultrasounds showed the position of the babies in the womb – doctors said they were arranged like a pyramid, the one closest to the cervix being nicknamed Hercules because he held all the others up. For more practical purposes, he also became known as "Baby A".

The decision was taken to deliver the babies at the start of the 31st week of pregnancy, less than the 40 weeks expected for a single birth, but at least three weeks beyond the point when doctors believed t he foetuses would survive. Multiple births often fail to go the full term, but doctors wanted Bobbi McCaughey's pregnancy to continue for as long as possible, since premature babies are most vulnerable to respiratory problems and feeding complications. With a 140-cm (55-in) waist, she could not hold out much longer and was desperate to bring her babies into the world. She remembers telling her husband: "Kenny, I just cannot be pregnant one more day." So when she began having contractions on the night of November 18, doctors decided to deliver the next day.

In the course of the pregnancy, medical staff at the centre had held numerous meetings to plan the approach to the birth and to anticipate any possible complications. There was understandable apprehension. With the previous septuplet birth in the United States, to a family in California, one of the babies was stillborn and three others died later. Also, in January 1997, severely underweight septuplets were born to a couple in Mexico. One was stillborn and the other six died later. More recently, in September 1997, a Saudi Arabian woman had given birth to septuplets, but only one had survived.

As many as 40 specialists were involved in the carefully organized McCaughey operation, supporting Dr Mahone and Dr Karen Drake who carried out the delivery itself. The first to be born, at 12:48 p.m., was Baby A. The family had already chosen names for each baby, and Kenny McCaughey was waiting to announce their arrivals from a slip of paper as he wanted one of the first things each child heard in their life to be their name being read out by their father. So, as soon as the first baby was delivered, Kenny announced him to the whole room as "Kenneth Robert". As Dr Mahone made each delivery, Dr Drake cut the umbilical cord and the infants were rushed out of the delivery room into the second operating room where tubes were attached to help them breathe.

Although the first baby seemed healthy enough, there was no time for complacency. Dr Mahone said: "After the first birth we needed to move quickly because once you've disrupted the blood flow to the uterus, you may be disrupting the blood flow to the other babies that are remaining."

Meanwhile Bobbi McCaughey was worried because the first two babies had made no sound as they were born, leading her to fear the worst. Then she breathed a huge sigh of relief when the third septuplet, Natalie Sue, was born yelling. "At least I knew that one of them was OK," she recounted. "In fact I cried longer than she did!"

The seven – four boys and three girls — were born in the space of just six minutes. Being the last born, Joel Steven, was at the highest risk of suffering from internal bleeding and was immediately declared to be in a critical condition. After doctors gave him a transfusion, however, his condition improved greatly. A relieved Dr Mahone told reporters: "As we delivered and saw the nice size of the babies and how vigorous they were, we were all very, very happy."

Kenneth Robert was the heaviest, at 3lb 4oz (1.871kg); Alexis May, 2lb 11oz (1.219kg); Natalie Sue, 2lb 10oz (1.191kg); Kelsey Ann, 2lb 5oz (1.49kg); Brandon James, 3lb 3oz (1.446kg); Nathan Roy, 2lb 14oz (1.304kg); and Joel Steven, 2lb 15oz (1.332kg).

It seemed that the whole of America was gripped by the story. The McCaugheys were even promised 30,000 free nappies plus a year of groceries. Kenny McCaughey's boss at a local Chevrolet garage announced that he was donating a new van, while neighbours in Carlisle promised to build a new house for the family. And there was no shortage of volunteers offering to provide round-the-clock support when the babies came home. The first of the septuplets, Kenneth Robert, was able to go home the following January.

Despite making medical history, it has not all been plain sailing for the McCaugheys. Two of the children, Alexis and Nathan, have cerebral palsy, as a result of which they have had difficulty walking. In 2004 Nathan underwent spinal surgery in the hope that it would be the first step toward him being able to walk properly on his own. Previously he could only walk for about 40ft (12.19m) without the aid of crutches. Inevitably the case of the McCaughey septuplets renewed the controversy about the ethics of fertility treatments, particularly those resulting in multiple births. Carl Weiner, director of the Center for Advanced Foetal Care at the University of Maryland School of Medicine, warned: "This is not the goal of fertility drugs. We should seek to avoid high-order multiple gestations whenever couples have a moral objection to selective foetal reduction."

Keyhole Surgery That Saved a Baby

Twenty minutes after being born three weeks prematurely, baby Ebony Martin stopped breathing and started to turn blue as she lay in her mother's arms. Medical staff rushed the baby through a series of tests to find out why she had stopped breathing and discovered that she had been born with a malformed oesophagus – it was not joined up properly. Instead of connecting her throat to her stomach, the upper oesophagus stopped at her neck and the lower part was attached to her windpipe. Consequently, she was unable to swallow and air was passing into her stomach. Without emergency surgery, she would have starved to death. It is a comparatively rare defect – affecting around one in every 5,000 newborn babies – and one that can invariably be corrected by conventional surgery. However, this would have left a large wound and extensive scarring, so when the 4lb 15oz (2.238kg) infant was transferred from St John's Hospital, Livingston, Scotland, where she had been born at the end of August 2001, to the Royal Hospital for Sick Children in Edinburgh, consultant paediatric surgeon Gordon MacKinlay decided to use pioneering keyhole throat surgery in the hope that the scarring could be restricted and the healing process accelerated.

ABOVE *Consultant paediatric surgeon, Gordon MacKinlay, used keyhole throat surgery to save baby Ebony Martin, who was born without the ability to swallow.*

The life-saving surgery was the first of its kind in Britain and, at two days old, Ebony was the youngest person in the world to undergo it. The keyhole surgery technique used involved passing three tiny tubes – each just 5mm (0.2in) in diameter – into the body. One tube carried a tiny camera and others bore miniature, needle-sized surgical instruments. The camera beamed images of the inside of Ebony's chest to a screen in the operating theatre where Mr MacKinlay manipulated the instruments while watching what he was doing on the screen. He disconnected the oesophagus from her windpipe and joined it back to the upper tube with fine stitches. Instead of the usual wide cut needed for the surgeon to access the chest, only the three 5mm (0.2in) holes were left to heal.

BELOW *Ebony Martin with her mother Alanna. Ebony was the youngest person in the world to undergo the pioneering keyhole throat surgery.*

Mr McKinlay said afterward:

"Performing this kind of keyhole surgery in a tiny baby is like operating inside a matchbox in comparison to an adult-sized chest, where the space is more like a shoebox. We are working with long instruments like knitting needles, which we're poking through these little tiny tubes going into the chest, and the most difficult part is actually stitching inside there. Stitching the gullet is amazingly difficult because there's so little room to manoeuvre, plus you've got the heart and lungs right beside it. We use conventional needles and suture material, but there is little room to turn the needle over and hand it from one instrument to the other. That was the most demanding bit and we found that we got terrible aching across our shoulders just from the tension of making sure that we put these fine sutures in very precisely and accurately."

Five days after the operation, an X-ray showed that the oesophagus was healing well and Ebony was now able to swallow milk. Two weeks after surgery, the oesophagus had healed completely and she was allowed home.

Ebony continued to make good progress and, as her mother Alanna testified, there was nothing wrong with her appetite. Thanks to the pioneering surgery, there is every chance that she can look forward to a healthy life.

The Birth That Beat Cancer

Seven years after being made infertile by chemotherapy, a Belgian cancer patient made history in 2004 by becoming the first woman to give birth following an ovary transplant. The birth of Tamara Touirat, achieved, according to doctors in Brussels, by replanting tissue from her mother's ovaries, has given fresh hope to those women who were afraid that they would never be able to have children in the wake of cancer treatment.

Women who undergo chemotherapy can lose their fertility as a result of the powerful drugs used in the cancer treatment, although in the case of young women the figure is put at less than 50 per cent. By contrast, if radiotherapy treatment affects the ovaries, it immediately causes infertility. For years doctors have been looking at ways of enabling cancer patients, whose fertility has been affected, to become pregnant. Teams in the United States and Europe had been freezing ovarian tissue with the aim of transplanting it back into the affected women, but while they had been able to restore normal ovarian function and periods, none had achieved a pregnancy until the Belgian case.

Ouarda Touirat was 25 when she was being treated for advanced Hodgkin's lymphoma in 1997. Prior to beginning the course of chemotherapy, a 0.04in (1mm) layer of tissue was removed from her left ovary, cut into sections and frozen in liquid nitrogen at a temperature of around minus 200 degrees

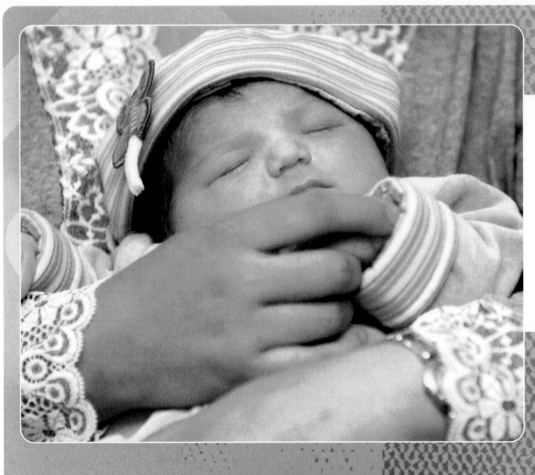

LEFT *The birth of Tamara Touirat, achieved by replanting tissue from her mother's ovaries, has given new hope to women who feared infertility following cancer treatment.*

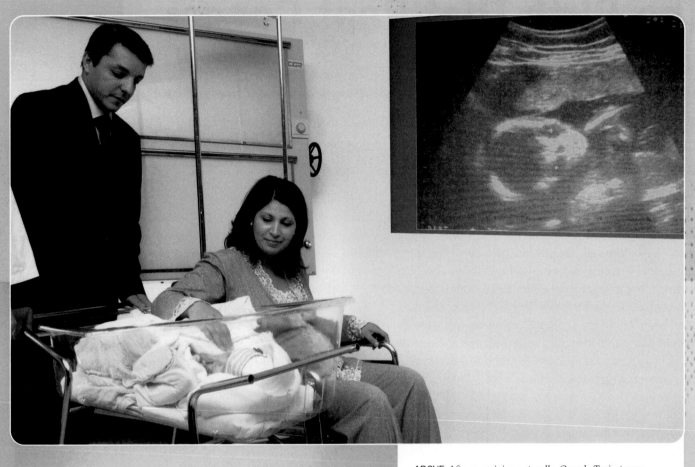

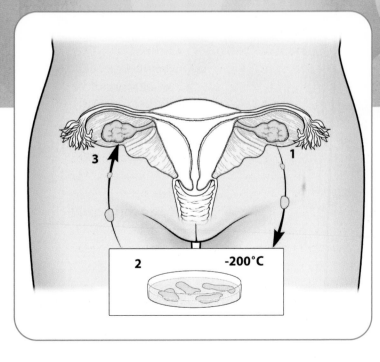

ABOVE *After conceiving naturally, Ouarda Touirat gave birth to 8lb 3oz Tamara at a Brussels hospital in September 2004. The baby's father, Malik Bouanati, admires the new arrival.*

LEFT *A layer of tissue was removed from Ouarda Touirat's left ovary, frozen in liquid nitrogen at minus 200 degrees Centigrade and replanted beneath her right ovary.*

Centigrade (minus 392 degrees Farenheit). After her cancer treatment, she stopped ovulating but in April 2003, once she was declared free of the cancer, her ovarian tissue was transplanted back into her body, immediately beneath her right ovary. Four months later, she was found to be menstruating and ovulating normally.

Eggs can be removed and used for in vitro fertilization (IVF). In this proceedure, an egg is surgically removed from the ovary and fertilized outside the body, in a laboratory. The resulting embryo is then implanted into the woman's uterus so that it can develop into a foetus. With Mrs Touirat, however, the tissue was placed at the ends of the fallopian tubes to allow a natural pregnancy. After conceiving naturally, she gave birth to 8lb 3oz (3.714kg) Tamara at Brussels' Cliniques Universitaires Saint-Luc in September 2004.

For Mrs Touirat, motherhood was beyond her wildest dreams. In 1997, she had only been married a year when she learned that she had cancer. Her dreams of starting a family appeared wrecked when she was told that, even if she did not die, there was a risk that the lifesaving chemotherapy treatment would bring on an early menopause and leave her infertile. So when she was offered a new technique, which, even though it had yet to succeed, might enable her to conceive normally, she jumped at the chance. When she was given the all-clear from cancer, however, doctors discovered that the menopause had already set in so they transplanted the ovarian tissue back into her body beneath the existing ovary.

Professor Jacques Donnez, who led the research into the new treatment at the Catholic University of Louvain, predicted that the birth would open new perspectives for young cancer patients facing premature ovarian failure. Many young women with cancer find the possibility that they may never have children almost as devastating as the news of their diagnosis. More than 2,000 teenage women are diagnosed with cancer each year. "It is a big message of hope for all women with cancer who have to go and have chemotherapy," said Professor Donnez. "This technique must be widely developed. It's really easy to freeze ovarian tissue. You just follow the protocol and wait. And the treatment is much less expensive than IVF. Every woman who receives chemotherapy for cancer should be offered the different options for fertility preservation. This is the way to go. Because of the progress made by medicine, more and more women are survivors of cancer."

Writing in *The Lancet*, the Belgian researchers said all the visual evidence suggested that the egg follicle, which ripened into the egg during the menstrual cycle when Mrs Touirat became pregnant, came from the transplanted tissue. A scan showed the development of an egg follicle at the site of the transplant, but nothing on either ovary in the menstrual cycle leading up to the pregnancy.

The news of the birth was greeted enthusiastically, but some scientists cast doubt on the Belgian claim, pointing out that it is not uncommon for ovarian function to recover of its own accord, and that therefore the pregnancy was not necessarily the result of the transplant. Dr Kutluk Oktay, of Cornell University in New York, who has produced embryos from an ovary graft, stated that Mrs Touirat had ovulated three times in the two years before the transplant, indicating that her ovaries had not stopped working completely. The risk of sterility for a woman of her age having her type of treatment is also only between 12 and 47 per cent. "I am cautiously optimistic," he said, "but there are parts of the research that need explaining before I'm 100 per cent convinced. We think there's no reason it can't be done, and it is very possible that this has worked, but we need more evidence to prove that it has."

The Belgian case was strengthened when Professor Donnez announced that a second woman had undergone a successful ovarian tissue transplant. The 28-year-old patient, who had ovarian tissue removed in 1999 before receiving radiotherapy for sickle-cell anaemia, started to menstruate again in January 2005, five months after the tissue was transplanted back into one of her non-functioning ovaries.

One person who is not too concerned as to the precise mechanics of the pregnancy is Ouarda Touirat. "I'm just very, very happy," she beamed, gazing lovingly at baby Tamara. "It's what I always wanted."

Although studies in this area have hitherto concentrated on cancer patients, women who want to extend their reproductive life could also use the technique. Women are born with a supply of eggs – usually about a million – that die off during their lives until they reach the menopause, when there are too few left to support a pregnancy. This new technique allows them to have some of their ovarian tissue removed, frozen, and stored. Once they have been through the menopause, it could be replanted, offering them another chance of motherhood. The prospect of freezing ovarian tissue for this reason has already aroused a fierce debate, partly because some view it as morally wrong (arguing that it goes against nature) and also because healthy women should think very carefully about whether they want to have an ovary removed. Every invasive operation carries a risk so, if the procedure does become widely available in years to come, any decision about having ovarian tissue removed and frozen for later use is certainly not one to be taken lightly.

The Birth of Triplets That Shocked Doctors

About one in every 100 pregnancies is ectopic, where the embryo starts to grow outside the womb, usually in the fallopian tube. With an ectopic pregnancy, the embryo can rupture the fallopian tube that invariably leads to the loss of the baby and massive internal bleeding, which can prove fatal for the mother. Because of the life-threatening nature of the condition, once an ectopic pregnancy is confirmed doctors recommend terminating the pregnancy – in fact the foetus is usually already dead. Occasionally, however, as with Ncise Cwayita's baby (pp.13–14), an ectopic pregnancy comes along that defies the medical textbooks. One such case was that of Jane Ingram and her baby Ronan.

Thirty-two-year-old Mrs Ingram, from Elmswell, Suffolk, UK had not been taking fertility drugs, but discovered 18 weeks into her pregnancy that she was carrying triplets. Ten weeks later, a routine scan revealed that one of the embryos had developed outside of the womb, in the fallopian tube. This lead doctors to believe that they were dealing with only a normal twin pregnancy. It then it became apparent that the third embryo had infact attached itself to the mother's uterus and created its own "womb" with its own blood supply in the mother's abdominal cavity.

Confronted with an extraordinary set of circumstances, the medical team at King's College Hospital in London faced a real challenge: to make sure that mother and children emerged safe and sound. The three were delivered 11 weeks premature in September 1999 by a 26-strong medical team, including three surgeons, three anaesthetists, three paediatricians, three midwives and 11 operating theatre staff. Doctors came in from leave to take part in the complicated operation, which staff described as a "military type procedure". After delivering the two girls in the womb by Caesarean section, the surgeons had to decide how to get the third baby, a boy, out from his home in his mother's abdomen. Because of his position, a Caesarean was not possible, so instead they moved Mrs Ingram's bowel aside in order to gain access to him and then cut open the sac that had developed around him.

Not only were all three born safely, but their weights – Olivia at 2lb 10oz (1.191kg), Mary at 2lb 4oz (1.021kg), and Ronan at 2lb 2oz (1.446kg) – were perfectly normal for triplets delivered at 29 weeks. They were all taken straight to intensive care where they were put into incubators and given caffeine to stimulate their breathing. There had been fears that Ronan, in particular, would experience breathing difficulties on account of his unusual method of development, but instead he appeared to have the strongest lungs.

Consultant paediatrician Dr Janet Rennie admitted:

> **"I was concerned that Ronan would have suffered stress from the position he was in and that he may have had problems breathing because he did not have the space and the nourishing atmosphere of the womb. But happily those concerns have not materialized. If anything he has been more robust than his sisters. He achieved natural respiration and came off the ventilator before the other two, perhaps because his lungs had to develop and were given the exercise they needed."**

Consultant obstetrician Dr Davor Jurkovic, who delivered Ronan and his sisters, said it was a miracle

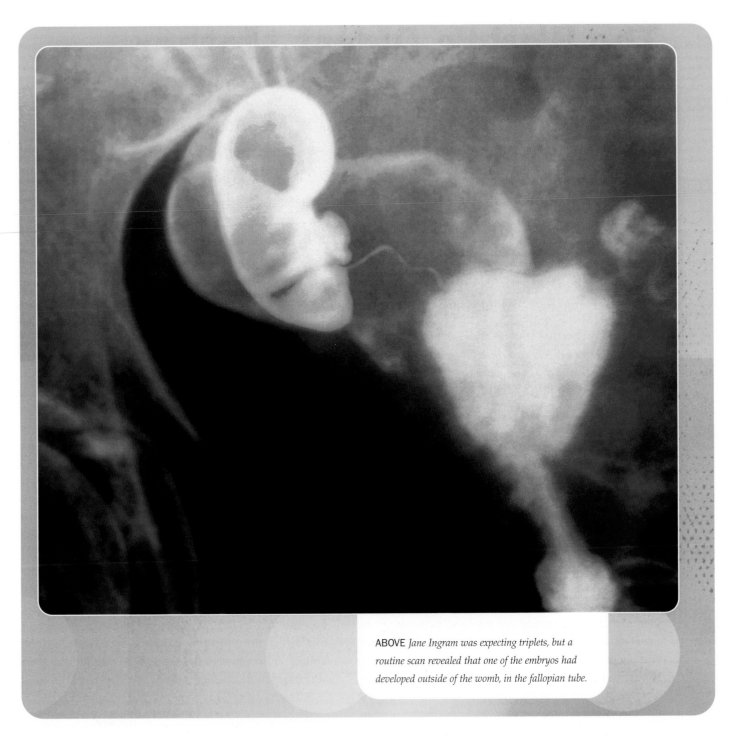

ABOVE *Jane Ingram was expecting triplets, but a routine scan revealed that one of the embryos had developed outside of the womb, in the fallopian tube.*

that the triplets and mother had survived. Indeed, he estimated that the chances of an embryo surviving in such circumstances were one in 60 million. He paid a glowing tribute to Mrs Ingram, adding that her bravery had contributed enormously to the success of the birth. "Jane was absolutely incredible," he told the BBC. "The attitude of the patient in situations like this is crucial and her attitude was always positive. She was fully aware of the risks but she was smiling all the way through, even in the last minutes before she was given an anaesthetic. Just before the babies were delivered, she turned to me and said, 'I have faith in you'".

There are fewer than 100 instances in the world of babies surviving outside the womb, but none of one baby surviving an ectopic pregnancy while two develop normally. In terms of multiple births, little Ronan Ingram had made medical history.

The World's Oldest Mother

"I feel like a normal woman, like any woman who has a child." So said Adriana Iliescu in January 2005, 48 hours after giving birth to a baby girl in the Romanian capital of Bucharest. However, Adriana Iliescu is not like any other mother – when she gave birth she was 66, making her the oldest mother in the world. The birth, which followed IVF treatment with donor eggs and sperm, has divided public opinion across the globe. To some she is a shining example that a woman should never give up hope of being able to have a baby; to others it is seen as a selfish act, one that is ethically unacceptable.

A retired university professor and author of children's books, Adriana had always wanted to have a child. She was married at 20 but because of the harsh living conditions under the country's Communist regime, she felt it was not the right time to start a family. She and her husband split up four years later and she embarked on an academic career. "I was a woman born to be a mother," she said. "Since I was a little girl, I have had this dream to have children. I always thought that after things changed politically here, I would try." But by the time she made the conscious decision to try for a child, it was too late for her to conceive naturally. Then in 1995 she heard about the success of IVF treatment. Adriana went to Italy that year to undergo tests but had to return to Romania because she could not afford to stay any longer. Back home, she contacted Professor Ioan Munteanu who had performed Romania's first IVF treatment that year. After having treatment for further delaying the onset of the menopause, she underwent fertilization treatments and in 2001 she fell pregnant for the first time. However, after four months the foetus died.

Undeterred by the fact that she was now reaching the age where most women become grandmothers rather than mothers, Adriana used her savings to fund a total of nine years of fertility treatment. She applied to Dr Bogdan Marinescu, director of the Panait Sirbu maternity ward in Bucharest, who, after using sperm and egg from anonymous donors, implanted three embryos in her womb. After ten weeks, however, one of the embryos stopped developing, leaving Adriana with twin girls. Then at 33 weeks, one of the remaining twins suffered the same fate, forcing doctors to operate earlier than intended. Accordingly, on January 16, 2005, at the Giulesti Maternity Hospital in Bucharest, Adriana Iliescu gave birth, by Caesarean section, to a 3lb 3oz (1.446kg) baby girl, Eliza Maria, born six weeks short of a full 40-week pregnancy. At 66, the new mother surpassed the record of a 65-year-old Indian woman, Satyabhama Mahapatra, who, in 2003, gave birth to a boy after being impregnated with an egg from her 26-year-old niece that had been fertilized by her neice's husband.

News of the Romanian birth sparked a huge debate over the ethics of women beyond the age of fertility having babies. Church leaders described it as "horrifying", "shocking", "grotesque" and "the ultimate act of selfishness", pointing out that Ms Iliescu will be 84 by the time her daughter is 18. The Equal Opportunities for Women Foundation also expressed disquiet, warning that the daughter might become frustrated in years to come because her mother is not like the mothers of other children. Young children can be cruel and might tease her about her mother's age.

Adriana was unprepared for the backlash. "I would have liked a traditional family," she admitted, "but I think a woman has to make a family however she can. It is simpler and more moral to do it scientifically. I think my way is more moral than having liaison after liaison in order to have a child. And it will be easier to tell Eliza how I did it by IVF rather than having to tell her some sordid tale about a lover, perhaps married, who left me. This way is in the modern spirit – a woman has another way of doing things." She was optimistic about her future, claiming that her family had a history of longevity.

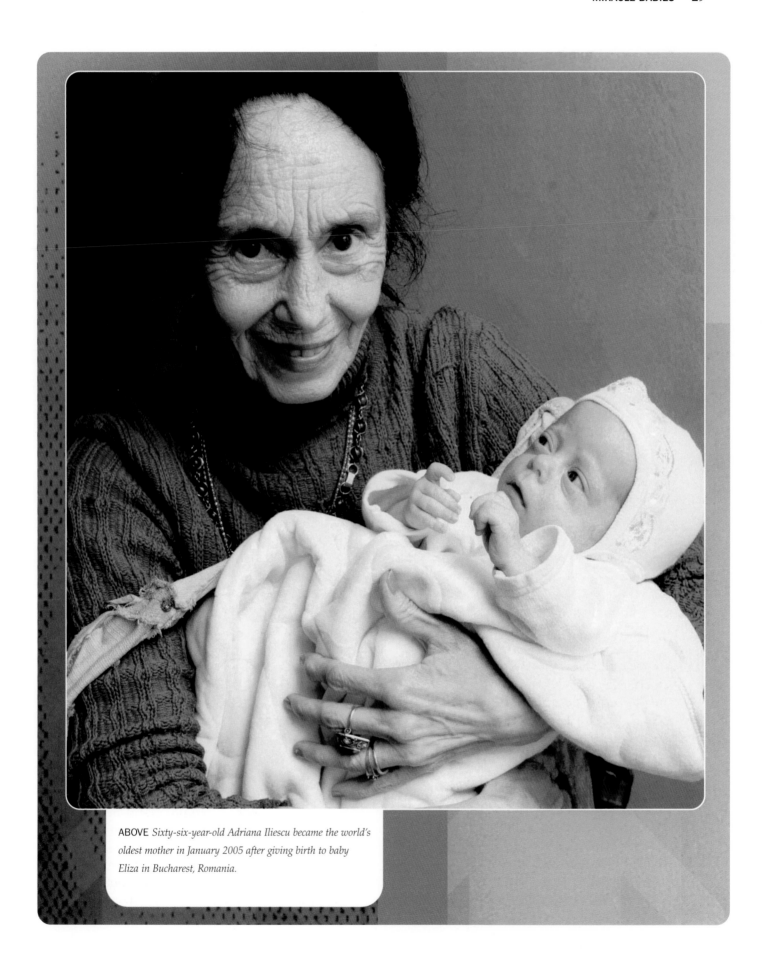

ABOVE *Sixty-six-year-old Adriana Iliescu became the world's oldest mother in January 2005 after giving birth to baby Eliza in Bucharest, Romania.*

Dr Marinescu defended the treatment, insisting that he had been impressed by Adriana's faith in God and her determination to have a child. He added that, despite her advanced years, she was in the right condition physically to carry a pregnancy. Fellow physicians were less convinced, stating that it was dangerous medically and morally, posing not only a threat to the mother's health, but also to the welfare of the child who deserves to have parents around to raise her. Underlining its unease, the Romanian Health Ministry said that it would not be encouraging artificial insemination of women who are no longer fertile. Although at the time of the Iliescu birth there was no law in Romania stipulating a maximum age for artificial insemination, there was a draft law awaiting approval by parliament that would ban fertility treatment for women who are over the normal reproductive age.

However, the reaction was not uniformly hostile. Some feminist groups applauded her decision and women in the street asked to touch her because they felt she was clearly loved by God. Adriana herself predicted: "I feel I have changed things. Young women will see me as courageous. In all this, one thing is clear: every person has a mission in life, and maybe that was my mission. That is why I succeeded in showing that women want and must have children."

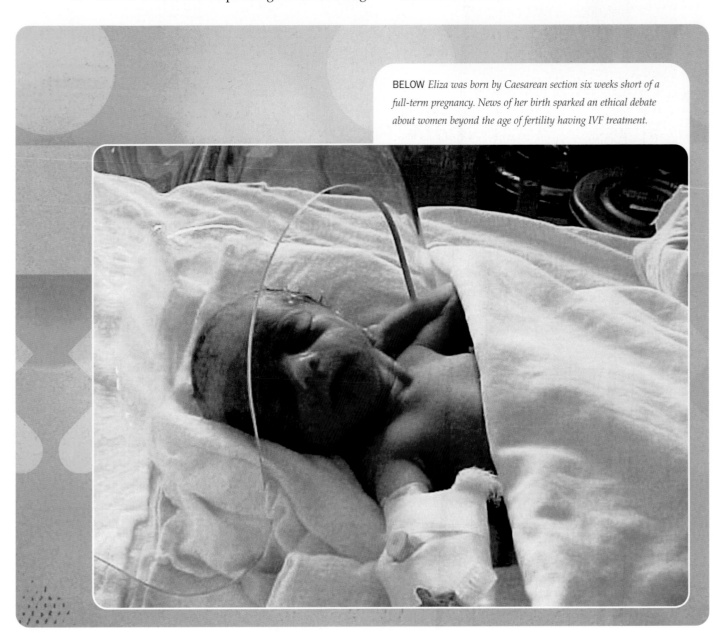

BELOW *Eliza was born by Caesarean section six weeks short of a full-term pregnancy. News of her birth sparked an ethical debate about women beyond the age of fertility having IVF treatment.*

ABOVE *Adriana Ilescu believes that the controversial birth of baby Eliza will offer fresh hope to other women desperate to have children.*

2 AGAINST ALL ODDS

For seven years a French woman by the name of Gabrielle Clauzel lay bedridden with chronic spondylitis, an inflammation of the joints between the vertebrae in the spine. Her 49-year-old body was frail and, with doctors unable to find a cure, the outlook was bleak. As complications set in, her very life was considered to be in danger. Then one day in August 1943 she was carried to a nearby church to attend Mass. Afterwards, to the amazement of everyone who witnessed it, she got up and returned home on foot. From that day on, she always enjoyed good health and never once suffered a relapse, going on to live to the ripe old age of 88. Five years after her startling recovery, a bishop declared her cure to have been miraculous. Indeed, despite the enormous progress that has been made in our understanding and treatment of illness, the occasional sensational recovery does appear to owe more to divine intervention than routine medicine. Lourdes has built its reputation on such cures. People who have been in comas for years suddenly regain consciousness; others regain their sight or their hearing for no apparent reason; and deadly tumours vanish without explanation. The medical profession is usually left baffled, but ultimately delighted by these patients who have fought back against all odds.

Back From the Dead

In July 1984, Terry Wallis, then 19, was riding with two friends in Arkansas when their pickup truck careered off a remote mountain road and plunged 25ft (7.6m) into a dried-up creek. The driver of the vehicle died; one passenger emerged without a scratch; and Terry was critically injured. He wasn't rescued until the following day and even then doctors warned his mother Angilee that her first-born son might not live more than a few hours, such was the extent of his brain-stem injury. Paralyzed from the neck down, he was left a quadriplegic and declared comatose. For weeks, months, years, he remained absolutely lifeless. Although his parents brought him home from his nursing home every other week and continued to talk to him, they had no idea whether he understood them. Apart from the occasional grunt or blink, that was how things stayed for 19 long years until suddenly, in June 2003, during a visit from his mother, he suddenly blurted out the word "Mom". This was the first word he had said since the crash. Not surprisingly, Angilee Wallis describes her son's return to consciousness as "a miracle".

Further words followed – "Pepsi", then "milk", then "Dad" – and before long the words had become phrases and short sentences. Soon he could say anything he wanted, although his speech remained slow and laboured. One of his first tasks was to catch up with his 19-year-old daughter Amber, born shortly before the crash. However, he remained in a time warp and asked to speak to his grandmother who had died several years earlier. Remarkably he was able to recite her phone number, something the rest of the family had long forgotten, and when asked who was President, he replied "Ronald Reagan". His family were only too happy to forgive the mistake. After all, they had never expected him to open his eyes again, let alone speak.

Theirs had been a long, painful journey. Angilee remembers that when she first saw Terry lying in hospital, she noticed his hand moving. She thought it was a good sign, only to learn that it was an indication of brain damage. Although doctors told her to prepare for a funeral, she never gave up hope. Once his condition had stabilized and he was moved to a nursing home in Mountain View in 1985, she drove the 50-mile return journey from home twice a week, year after year, to visit her son. On home visits, she would wheel him around to familiar places, to meet familiar people, and she would talk to him and read to him constantly, in a bid to elicit a response. "I wanted to keep on talking to him and taking him home," she says. "It just got to be routine. It is what we did, but I can't say I really thought he would get better."

Then, on one of Angilee's routine visits, the nursing home worker escorted her to Terry's room and, as she always did, asked her patient who his visitor was. "He just said 'Mom'," recalls Angilee proudly. "I just fell over on the floor!" Not only were she and the nursing home worker surprised, so was Terry.

"You could tell by the look on his face," says Angilee. "His eyes were kind of big."

To understand Terry's state of mind since 1984, it is necessary to know a little about the brain itself. The brain is an extremely delicate organ. Any jolt causes it to slide around, compressing and expanding as it goes. In the event of a severe jolt – such as Terry's car crash – the billions of neurons (nerve cells) that make up the brain are stretched, twisted and sheared. Whereas the penetration of the skull by an external object – such as an iron railing – may look more

ABOVE *Terry Wallis pictured with his wife Sandra four months before the car accident that left him in a coma for 19 years.*

RIGHT *In June 2003, during a visit from his mother Angilee, car crash victim Terry Wallis said his first word in 19 years. It was "Mom".*

serious, the impact in such instances is often localized. By contrast, although the skin may remain unbroken after a car smash, the impact on the brain can be more diffuse and therefore more catastrophic. The brain will be thrown backward and forward against the base of the skull, causing extensive damage at the various points of impact. Head injuries are often accompanied by a loss of consciousness. In mild cases, this takes the form of concussion, lasting no more than a few minutes. Concussion is caused by temporary neuronal paralysis, but there is no damage to the actual brain. Comas are more difficult to define, but the term is generally used to describe someone with their eyes closed all the time and who is unable to communicate or respond to instructions. Comas are typically caused by damage to the brain stem, an arousal centre situated at the base of the brain, the injury effectively shuts down consciousness. Nobody is able to explain fully why people suddenly wake from long comas. Terry's doctors said that the constant talks from his mother helped to keep his mind going, but in any case the lack of responsiveness during a coma is a result of damage to either the brain stem, that controls the body's level of arousal, or its connections with the other senses. The rest of the body may remain unaffected and function normally. You do not need to be conscious to breathe, produce saliva or digest and excrete food. All of these functions are controlled by the autonomic nervous system, which is under the influence of the part of the brain known as the hypothalamus. In fact many people recovering from comas say that they were fully aware of everything going on around them, but were simply unable to communicate.

Nevertheless, the timing of Terry's recovery was astonishing. "It's kind of peculiar," mused his father Jerry. "He wrecked on Friday the 13th and 19 years later he started talking on Friday the 13th."

One unexpected side effect was that Terry sometimes started talking dirty. When the speech therapist asked him what she could do for him, he told her: "Make love to me." On another occasion, when asked how he felt, he replied "Horny". "That's kind of strange," said his father, "because he would not have talked dirty before he wrecked." But it's a small price to pay for having his son back and, although he remains paralyzed from the neck down and still has problems with his short-term memory, there is no escaping the fact that Terry Wallis really has come back from the dead.

The Woman Whose Skin Fell Off

When a young American woman developed a severe allergic reaction to an antibiotic drug, the skin on her body started peeling off in sheets. Doctors held out little hope of her survival, but amazingly, just three weeks later, she was back home and well on the road to a full recovery.

At the start of December 2003, 29-year-old Sarah Yeargain from San Diego, California, was put on a ten-day course of the common antibiotic Bactrim to clear up a sinus infection. She had just finished the course when she started to get some minor swelling and discoloration in her face. It then progressed into blistering on her lips and swelling on her eyes before the blisters began spreading all over her face, chest and arms. She consulted a doctor who simply recommended taking paracetamol for the increasing pain and advised rest and recuperation. By the next day, however, all the skin had peeled off her feet and turned to huge blisters that, when they popped, started to ooze. Consequently, she was unable to walk and had to be lifted out of the house by her mother, Katherine, who could see her daughter's skin coming away in her hands as she carried her to the car to take her to hospital.

By the next day Sarah's skin was coming off all over her body, including the skin on her internal organs and membranes covering her mouth, throat and eyes. She lost skin from her entire scalp and her hair fell out. Doctors at the University of California Regional Burn Centre in San Diego told Katherine

Yeargain that there was little chance of her daughter surviving. "Generally with 100 per cent sloughing there is a 100 per cent mortality," said Mrs Yeargain, "but we just prayed. It was horrible to see her lying in that hospital bed without any skin."

Sarah had experienced a rare and severe allergic reaction called toxic epidermal necrolysis, where the body's immune system malfunctions after it is exposed to a drug, in this case Bactrim. All of the skin comes off because victims experience a total body reaction. Furthermore, the loss of skin causes fluids and salts to ooze from the raw, damaged areas, which can easily become infected. Dr Daniel Lozano from the Burns Centre said: "Once the skin starts to slough, there's no stopping it. It's rather dramatic to see it coming off in sheets." Dr Lozano and his colleagues treated Sarah by covering her entire body with a skin substitute called TransCyte, which was developed at the specialist unit. Applied with staples over a 48-hour period, it created a seal to prevent infection and helped Sarah's own skin to heal.

Her grandmother, Marjorie Yeargain, said of Sarah's condition:

"I can only compare it to blisters. When you pop a blister, you are left with raw skin underneath, which is bright red. That is what Sarah looked like, except it covered the whole of her body. I cannot imagine the pain. Fortunately she was given sedatives that dulled the pain and kept her unconscious because the doctors were worried she would have a heart attack if she had to endure the agony. She was also given a drug that induces some form of amnesia. They suggested this because the skin loss was so traumatic for Sarah. It was touch and go for a while. We honestly thought that we were going to lose her. "

The doctors also gave Sarah drugs to prevent internal bleeding and within a week her own skin was back. After a few weeks, the TransCyte started to dissolve, allowing the new skin to take over. It was anticipated that in time she would make a full recovery, the new skin being just as strong and durable as the skin that she lost. There wasn't expected to be any scarring because, unlike with burns, only the top layer of skin was damaged.

Sarah Yeargain is thought to be the first person to have survived a full attack of toxic epidermal necrolysis. Her family and some of the medical staff called it a miracle. As one nurse remarked: "I think with the magnitude of the skin loss she had, that there was a divine hand in her recovery."

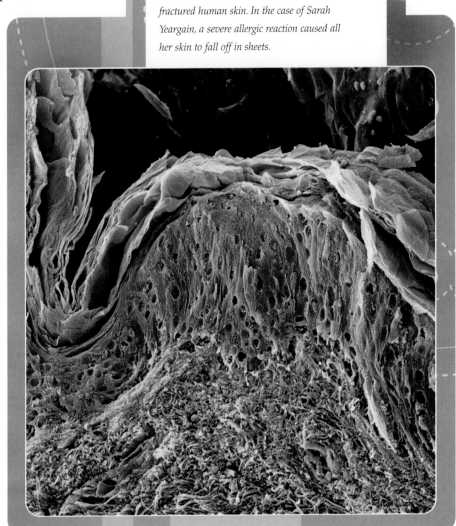

BELOW *A scanning electron micrograph of freeze-fractured human skin. In the case of Sarah Yeargain, a severe allergic reaction caused all her skin to fall off in sheets.*

Sight Regained After a Bump on the Head

Lisa Reid lost her sight after a cancerous brain tumour cut the blood supply to her eyes. Although a delicate operation succeeded in removing the tumour and saving her life, her optic nerves remained permanently damaged. By the time she was 14, she was left completely blind with no prospect of ever being able to see again. Ten years later, however, her sight miraculously returned following a bump on the head. Her case has left doctors baffled.

The cancer that robbed Lisa of her sight was diagnosed when she was just 11 years of age. After discovering a large brain tumour, doctors put her chances of survival at no more than five per cent. Radiology and an operation to remove the cancer were successful, but the tumour had already damaged her eyes, cutting off blood and putting pressure on her optic nerves. Three years later, she

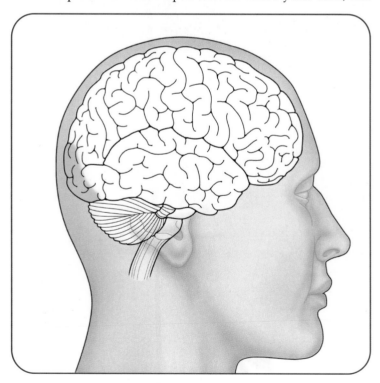

ABOVE *The occipital lobe is the area of the brain responsible for vision. A brain tumour had damaged Lisa Reid's eyes, cutting off blood and putting pressure on her optic nerves.*

was declared legally and permanently blind, her eyes able to detect only light and dark. "Everything was against me," she admitted, "but I'm a stubborn old redhead."

Throughout her ordeal she never gave up hope that she might one day see again. She maintained the strength to remain active by teaching others about blindness and by raising money for the organization that trained her beloved guide dog, a Labrador Retriever named Ami. It was Ami who unwittingly played a part in Lisa's sudden change of fortune.

On the evening of November 16, 2000, 24-year-old Lisa bent down to kiss Ami goodnight in her home near Auckland, New Zealand, but as she did so she hit her head hard on the side of a coffee table. "I kind of lost my balance," she said. "I hit my head on the floor and coffee table at the same time." When she woke the next morning, she found to her amazement that she could see for the first time in ten years. "I just saw the white of my ceiling. Looking around my room… the light shining through the curtains… the window frame… wow, the colours… Looking at Ami, she's just beautiful."

Lisa decided to keep the news to herself for a few hours, preferring instead to play with Ami in the back yard. It was not until later in the day that she contacted her family, and read a health warning from a cigarette packet to her mother over the telephone. Her mother, Louise, remembered: "Lisa called me and said, 'There's been a change. Listen to this.' Then she started reading to me. I was completely blown away."

Still unsure whether her sight would last, Lisa waited until the following day before throwing away her walking cane and telling more people. When friends and family turned up on her doorstep to celebrate, she could not identify them by sight. Her brother had changed from a 12-year-old boy into a man and it was the first time she had seen her boyfriend of two months.

Doctors could find no explanation for Lisa regaining her sight. The optic nerve is one of the few human tissues that cannot regenerate itself, and subsequent tests showed that hers is still as damaged as it was before. Auckland Hospital ophthalmologist Ross McKay revealed that Lisa had regained 80 per cent vision in her left eye, although her colour vision remained limited. "In my 25 years as an eye specialist I have never encountered a similar case. For some reason she's got her sight back, but I don't know how. And I don't believe in miracles."

Lisa is aware that her vision could disappear again as suddenly as it returned, but she remains unfazed. "The doctor who told me I would never see again was the same one who said my vision was now 80 per cent.

To be able to look him in the eye and see him when he told me that, it was pretty good. And if my vision went, just like that, I would continue to feel gifted and happy because I would still have experienced a miracle – an act of God – and still have a special experience to share with everyone."

Her boyfriend added: "She's such a strong and passionate woman. There was no way those eyes were never going to see again. You could just tell they would."

It seems that although Lisa Reid's astonishing recovery confounded medical science, it almost came as less of a surprise to Lisa herself and to those closest to her.

There are a handful of precedents for sight returning following a sudden blow or shock. Eighty-four-year-old Ellen Head had been 90 per cent blind for three years when her flat at Newcastle, Australia, was shaken by an earthquake for five seconds in 1989. Following the tremor, she found that she had regained her sight.

LEFT *Lisa Reid swimming with her beloved guide dog Ami in New Zealand. Lisa describes regaining her sight after a bump on the head as "a miracle – an act of God".*

The Miracles of Lourdes

Each year six million people make the pilgrimage to Lourdes in the French Pyrenees to take a dip in the waters that are claimed to heal cases beyond the powers of medical science. Ever since the Virgin Mary was said to have appeared before a poor peasant girl, Bernadette Soubirous, in 1858, the town has become a religious shrine and in the ensuing years the Roman Catholic Church has officially recognised over 60 miracles. Hundreds more visitors maintain that the waters of Lourdes have succeeded where doctors had failed.

Jean-Pierre Bély, a 51-year-old father of two from western France, had been stricken with a severe form of multiple sclerosis for 15 years. Confined to a wheelchair, he had officially been declared an invalid by the French social security system. Then, in 1987, he went on a pilgrimage to Lourdes. While lying in the sick room, he was suddenly overcome by a powerful sense of inner liberation and peace and heard a voice saying, "Get up and walk". He recalls: "A feeling of cold, getting stronger and stronger, invaded me. Then there was a sensation of heat, slight at first and then difficult to bear." He sat up, hung his legs over the side of the bed and took his first steps for three years. "I felt like a child learning to walk."

At the end of the pilgrimage, Bély travelled back to the station in his wheelchair so as not to appear different from his "companions in sickness". But he was able to board the train unaided and by the time he had reached his home town of Angoulême he had regained full use of his physical faculties. Subsequent medical reports showed that there was no trace of the disease and leading multiple sclerosis experts admitted: "Such a cure is not just unusual but inexplicable".

Another of the stories held up by the Church as proof of the power of Lourdes is that of a young Parisian woman, Louise Jamain, who, at the age of 22, was dying of tuberculosis, a disease that had already killed her mother and four of her five brothers. She had been hospitalized for seven years and, spitting blood and running a temperature of 40 degrees Centigrade (104 degrees Farenheit), was given a matter of days to live when she decided to visit Lourdes in the spring of 1937. Those closest to her strongly advised against it, fearing that she would not return alive, and their forebodings appeared justified when, following a dreadful journey, she fell into a coma and the last rites were administered. Nobody expected her to survive but two days later she woke at three o'clock in the morning, sat up in bed and suddenly announced that she was hungry. "I kept thinking about haricot beans," she recalled. "What I really wanted was a big plate of them. A nurse told me to get up and walk, but I doubted at first. She had to insist before I tried. Then I found I could stand up on my own." Medical examinations showed that the tuberculosis, for which there was no known cure at the time, had simply disappeared. She quickly put on weight and two months later landed her first job – at a printer's in Paris. In the 1940s she married and had two children, and in 1951 the Church declared her case to be a miracle. She has gone on to live well into her 80s without suffering any relapse.

Three-year-old Frances Burnes was given just weeks to live in the 1970s after surgeons had diagnosed malignant cancer. As a last resort, her mother, Deirdre, flew her to Lourdes. There, Frances bathed in the waters and a few days later returned to her home in Glasgow where she amazed doctors by beginning to make a remarkable recovery. Within three weeks, the doctors could find no trace of the carcinoma, which had riddled her with so much pain, and a month later she was back at playschool. One specialist acknowledged: "In medical terms, we can only call it a miracle".

Lydia Brosse had undergone several operations for bone disorders and abscesses that left her exhausted, emaciated and, due to intestinal and nasal haemorrhages, anaemic. In this pitiful state, the 41-year-old was taken to Lourdes in 1930 in the hope of achieving a cure, but here was no obvious improvement during her stay. It was only on the return train journey from Lourdes to Saint-Raphael,

RIGHT *A blessing at Lourdes. After one such visit, multiple sclerosis sufferer, Jean-Pierre Bély, suddenly found he could walk after 15 years confined to a wheelchair.*

under the eyes of a doctor, that she suddenly found the strength to get to her feet. Furthermore, her fistulas were found to be closed. The following day her doctor certified "a flourishing state of health, all wounds healed, and discharge had disappeared". Three months later, with no further haemorrhaging, she had put on weight and was a picture of health. She eventually died in 1984 at the ripe old age of 95.

In 1940, when she was 17, Yvonne Fournier was badly injured in an accident at work. Her upper left arm had become entangled in the driving belt of a machine and the incident left her traumatized and in unbearable pain. Her arm was virtually useless. Over the next five years, leading specialists in the field carried out nine operations, but they achieved nothing more than slight, temporary relief. Meanwhile, Mlle Fournier was awarded a pension equivalent to that for an amputated limb. Then in August 1945, after the Second World War, she went on

the first pilgrimage to Lourdes. Following a dip in the waters, she felt her left arm return to normal. Amazingly, power and movement were restored. In 1959 the International Medical Committee ruled that her cure was "…instantaneous and definitive. Moreover, it is medically inexplicable". In the same year, the Church recognized the cure as miraculous.

Devastating injuries sustained in the First World War left Liverpudlian Jack Traynor with a total-disability pension. He had received two bullet wounds, one of which made a hole in his skull, the other rendered his right arm paralyzed. His health deteriorated further when, in 1923, he began to suffer from epilepsy and was unable to walk. That year he was taken on a pilgrimage to Lourdes where he was lowered carefully into the communal bath. Four days later he jumped out of bed, washed and shaved himself and walked out of the hospice unaided. Returning to England, he found a job as a coal

ABOVE *Six million people make the pilgrimage to Lourdes each year to take the holy waters that are said to possess extraordinary healing properties.*

merchant, got married, fathered two children and led a perfectly normal life until he died from pneumonia in 1943. The Ministry of Pensions continued to pay him the full disability pension because they refused to accept that someone who had been classed as completely disabled could become totally cured.

At the age of three, a French boy, Francis Pascal, developed meningitis. Although he did not die from it, he suffered paralysis of the lower limbs and to a lesser degree in the upper limbs, and loss of vision. At least a dozen doctors ruled out any hope of a cure but the following year (1938) he was taken to Lourdes and immediately recovered his sight and lost his paralyses. When he returned home, he was examined again by some of the doctors who had previously seen him and in 1946, it was confirmed that he had been inexplicably cured.

In 1962 Italian soldier Vittorio Micheli was admitted to hospital in Verona for treatment on his left hip. After conducting various tests, doctors carried out a biopsy that revealed the presence of a malignant tumour. He remained in hospital for another year, during which time his condition deteriorated steadily until his hip joint was totally destroyed. Then, in the summer of 1963, he went to Lourdes where he bathed, despite being encased from pelvis to foot in a plaster cast. Within a few weeks his hip had reconstructed itself, he was able to walk and the pain had vanished. Thirteen years later, with no sign of a relapse, his cure was declared a divine miracle.

Delizia Cirolli was born in Sicily in 1964, the eldest of four children. At the age of 11, she began to feel persistent pains in her right knee, which turned out to be caused by a malignant tumour. The surgeon proposed amputating the entire leg, but when her parents protested he prescribed a course of radiotherapy instead. However, the poor girl was so traumatized by the prospect that she had to be taken home and no treatment was administered. Touched by her pupil's suffering, her schoolteacher suggested sending her to Lourdes, and the girl and her mother made a pilgrimage in August 1976. It proved a painful experience and when Delizia returned home, she was no better – indeed X-rays revealed that her condition had worsened. Nevertheless, her family continued to pray for her recovery and she had constant access to water from Lourdes. In mid-December Delizia stopped eating and her distraught mother started making the mortuary dress in which, under Sicilian tradition, her daughter would be clothed after her death. Then just before Christmas, when all seemed lost, Delizia suddenly felt better and was able to get out of bed without help and walk around. Her appetite returned and she was completely cured, a fact recognized by the Church in 1989 when the Archbishop of Catania declared her recovery to be "…scientifically inexplicable: a miracle".

The recoveries witnessed at Lourdes may well be the work of God or, of course, they may represent the ultimate in self-suggestion – proof that if you believe firmly enough in your recovery, it can happen. Indeed, it has been found that one in 60,000 cancer patients, believers or otherwise, gets better for no apparent reason. Whatever the truth about the healing powers of Lourdes, the faithful like to believe in miracles. And why not?

ABOVE *Sicilian girl Delizia Cirolli faced the prospect of losing her right leg to a malignant tumour until the waters of Lourdes brought about a miraculous cure.*

Mystery Accident Victim Recovered Speech

Nearly six years after being seriously injured in a car accident, a Brazilian man was finally able to tell doctors who he was because he suddenly regained the power of speech. In December 1997, Amauri Calixto was hit by a car while crossing the highway near Curitiba, an accident in which one of his children was killed. The accident left him partially paralysed in the arms and legs and with cranial injuries that caused him to lose both his memory and the ability to speak. Because he was carrying no documents with him when he was admitted, doctors at Cajura Hospital in Curitiba had no idea of his name or where he lived. Appeals regarding the mystery patient's identity were made on television and in newspapers but to no avail. Nobody seemed to know or recognise him.

Following emergency surgery and a lengthy stay in intensive care, he was transferred to the neurology department. By then, doctors had given up all hope of ever identifying him. However, they continued to give him a daily session of physiotherapy in the hope of improving his mobility and started him on a course of speech therapy. During one of these speech sessions, on September 30, 2003, he uttered his first word in almost six years. The nurses were amazed, although there was still no indication that his memory had returned and therefore they were no nearer discovering his identity. That breakthrough came shortly afterward when he revealed his name to the nurse who was giving him his bath. Over the next couple of weeks, he steadily recovered his speech and was able to give details of his home and family. The hospital wasted no time in contacting them.

The news that Amauri had been in hospital since 1997 came as a shock to his sister Rosalba who said she thought her brother was dead. "All this time had passed, we lost our hope," she said. "But now he will come home and we will take care of him."

One of the doctors treating him, Dr Beatriz Alves de Souza, admitted that he was pleasantly surprised by the turnaround in his patient's condition. "We realise now that he didn't lose his hearing and because of that he was able to recover his speech. By his example we can tell that nothing is impossible."

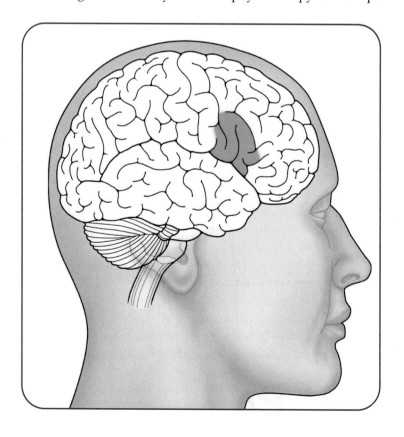

LEFT *In the human brain, speech is controlled from an area in the frontal lobe of the dominant hemisphere. Cranial injuries from a road accident left Amauri Calixto unable to speak for nearly six years.*

Hearing Returned with Pregnancy News

In April 2004, 21-year-old Emma Hassell went to the bathroom to take a shower. Suddenly her ears "popped" and her world fell silent. She remained deaf for the next seven months until, on the same day she had been told that she was pregnant, her hearing returned just as unexpectedly as it had vanished.

Emma's distressing experience began on what should have been a joyous occasion – the day she celebrated her engagement to boyfriend Kevin Love with a night out. Before getting ready, the Southampton care worker went upstairs for a shower, but 20 minutes later she found herself standing in the bathroom, unable to hear a sound. Furthermore, she had no idea what had happened in the intervening period.

> **"I was just about to have a shower," she said, "and then everything went muffled and very faint, and then it went completely. At first it just felt like my ears had popped, but it didn't pop clear – instead it popped and went completely. I just couldn't understand what was going on. I remember shouting down to my mum that I couldn't hear, but I don't know what happened. I just lost about 20 minutes. We thought I must have hit my head, but nothing showed up."**

She was taken to Southampton General Hospital, where tests confirmed that she had suffered a complete loss of hearing. Doctors were unable to explain why she had suddenly gone deaf but suggested that it could have been psychological. So Emma went to see a hypnotherapist who practised Emotional Freedom Techniques (EFT), an emotional form of acupuncture using fingertips instead of needles to stimulate energy points around the body. She had about eight sessions of EFT.

Then, on the morning of November 1, she did a home pregnancy test, which came out positive. This was wonderful news for Emma because she had

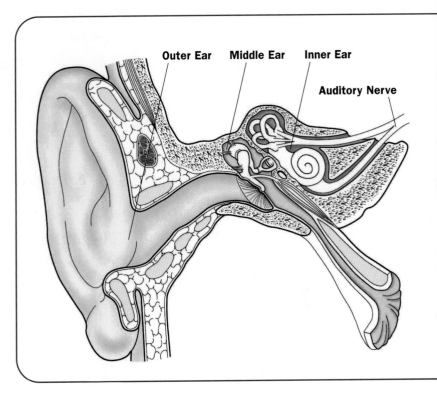

ABOVE *The anatomy of the human ear. Doctors were baffled by Emma Hassell's sudden and complete loss of hearing.*

suffered a miscarriage in 2002 and had been warned that she may not be able to conceive again. She was still in a state of shock when, a couple of hours later, she sat down to watch television.

"I was sitting on the couch watching 'Will and Grace'. I had the subtitles on and was concentrating on their mouths to read their lips, but then I thought that I could hear them. I was worried my mind was playing tricks on me. I tried tapping my fingers to see if I could hear that and then I phoned Kev to see if I could hear it ringing. I could, but I was so shocked that I hung up. I was worried and I was thinking, 'Please let this be real'. I phoned Kev back and he was speechless. I told him that it was not the time to have nothing to say! I just wanted him to keep talking so that I could start to believe it was real."

"It was just so unexpected, although the hope was there. I hadn't given up hope, but was feeling less like it would return. It's very, very strange."

Although Emma remains as mystified as the professionals as to why her hearing disappeared and returned, she is convinced that it was psychological. There seems no obvious explanation as to why she lost her hearing in the first place, but could its reinstatement be connected in some way to the euphoria of discovering that she was unexpectedly pregnant?

RIGHT *Emma Hassell and her fiancé Kevin Love. After seven months of deafness, she amazingly regained her hearing on learning that she was pregnant.*

The Man Who Learned to See

When Mike May was three, a jar of fuel for a miner's lantern exploded in his face, destroying his left eye and scarring the cornea of his right. Defying his blindness over the next 43 years, he led a full and active life, playing a variety of sports, earning a master's degree in international affairs at university, landing a job with the CIA, and becoming president of a company that makes talking global positioning systems for the blind. Along the way he found time to help develop the first laser turntable, marry, have two children and buy a house in California. During that time he had some ability to perceive light, but could not make out form or contrast. "Someone once asked me if I could have vision or fly to the Moon, what would I choose," he once wrote. "No question – I would fly to the Moon. Lots of people have sight; few have gone to the Moon."

Then, in November 1999, he came back to his senses. At St Mary's Hospital, San Francisco, surgeon Daniel Goodman dropped a doughnut shape of corneal stem cells on to May's right eye (his left eye was too severely damaged to be repaired). The cornea is the clear part of the outer layer of the eye that covers the iris and the pupil. It allows light into the eye and refracts the light rays to help the lens focus them upon the retina. Put simply, it is the eye's front window. The stem cells replaced scar tissue and rebuilt the ocular surface, preparing the eye for the grafting of a new cornea, and formed a protective layer over his new cornea to prevent clouding.

On March 7, 2000, when the wraps were removed, May got his first ever look at his wife, his children. "I know you're smiling," he said to his wife shortly afterward, "because your cheeks went up, and cheeks go up when someone smiles."

But operations of sight restoration – particularly with long-term blindness – are notorious for their psychological side effects. Some patients are left wishing they were still blind. Convinced by well-wishers that vision will offer a new appreciation of the world, they instead find themselves living in dread of performing simple everyday actions, such as walking down stairs or crossing the road. Dispirited and depressed, a sizeable proportion revert to the world of the blind, preferring dark rooms and walking with their eyes shut. Before Midlands shoemaker Sidney Bradford had his sight restored in

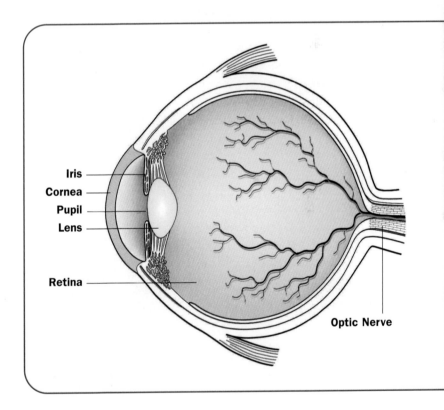

ABOVE *The cornea is the eye's front window. It allows light into the eye and refracts the light rays to help the lens focus them upon the retina.*

1959, he thought the world was heaven, but when he was able to see, he became distressed by life's little imperfections, such as chipped paint. He hated how his wife looked and could not distinguish between facial expressions. The actual images of things, such as an elephant, competed with his mind's eye picture of them as "an animal with a tail at each end". As a blind man he had been successful and competent, but now with sight he felt inadequate and entered a spiral of depression. "Sight restoration is a dangerous gift," says psychologist Oliver Sacks. "There is a dilemma in doing it and there are examples of people just closing their eyes on the world again."

ABOVE *When the bandages were removed following eye surgery, Mike May, who had been blind since the age of three, got his first look at his wife and children.*

In the last 200 years there have only been around 20 cases where people have had their sight restored after being blind since childhood and most of those still had slightly flawed corneas after surgery. To work properly, the cornea must be crystal clear – when Dr Goodman peered into Mike May's eye after the transplant, he saw a lens that ought to provide perfect vision. May had zero expectations of what would happen when the bandages came off and was pleasantly surprised to find that a certain amount of vision had been restored, but there was a problem. Although his optical hardware was up to standard, his brain had never been programmed to process the visual information it was receiving. Following his first business conference since the transplant he wrote: "I found it very distracting to look at people's faces when I was having a conversation. I can see their lips moving, eyelashes flickering, head nodding and hands gesturing. At first, I tried looking down, but if it was a woman in a low-cut top that would be even more distracting."

May holds the world speed record for downhill skiing by a blind person. In his competitive days he used to slalom down the slopes at 65mph, steered by a guide positioned ten feet ahead to shout "left" and "right". Six weeks after the removal of the bandages, Dr Goodman allowed him to go skiing again, and May took his family up to the Kirkwood Mountain Resort in the Sierra Nevadas – the place where he had first learned to ski and where he had later met his wife Jennifer.

The sun was shining, the trees were green (a deeper green and much taller than he had imagined), and the slopes were surrounded by beautiful cliffs, although he could not be sure whether they were a couple of miles away, as logic dictated, or the couple of 100 yards as they appeared. "Seeing snow for the first time on a sunny day was the most exciting visual thing," he said. "Also the beauty of the mountains, and I was so struck by the magnificence of trees. To see how they went up and up, till they nearly fell backward, was just extraordinary. The world is incredible and so new yet so familiar." With only one working eye, he already lacked depth perception, but he also had little experience of reading the shades and contours of a landscape. Skiing down the mountain, he struggled to distinguish shadows from people, poles, or rocks. At first he tried to work out the lie of the land scientifically: if a certain slope was being lit from the side and a shadow fell in such a way, then the slope must be convex. But once he hit his first bump, he was tempted to close his eyes and ski the way he knew best.

On being tested five months after surgery, May was able to perceive slight movements of a bar and recognise simple shapes. Eighteen months later, he could make out form, colour and motion almost normally, but could only identify about a quarter of everyday objects shown to him. Facial recognition was particularly problematic. All human faces looked alike – even those of his family. "I can't process a face," he said in 2002. "I can't tell the difference between a man's face or a woman's face either. There are clues, like long hair or earrings, but those distinctions are blurred these days so I go with plucked eyebrows as the best guide." The visual capacity of his brain remained that of a toddler. "Basically I'm an articulate three-year-old," added May who continued to think of himself as a blind man, since he was still using a cane and learning how to see. "When I need to concentrate, when the flow of information is too much, I close my eyes to stop the confusion."

His perception of motion was found to be his strongest visual faculty. Although he could barely recognise a motionless ball, he became adept at catching a moving one, and playing ball with his young sons became one of his greatest pleasures. However, his newly acquired low level vision continued to have a downside, not least when it came to skiing. "All of a sudden there's all this information flying in distracting me, making me tense up. In skiing you don't want to do that... I was falling all over the place." He was also nervous about crossing the road, something he did confidently when blind.

In Mike May's remarkable journey it has often been the ordinary things that have fascinated him the most. "One day I saw beautiful sparkling lights in the air in front of me," he remembers. "They were so bright and fleeting, I asked what they were. It was dust, and it gave me a whole new concept of what dust was." The excitement of seeing new and wonderful things each day means that, despite the struggles, he heartily recommends the stem cell transplant surgery. "This type of operation may work only for a small percentage of other blind people. Life is very full without seeing. But if the opportunity comes along, check it out."

Deadly Tumour Vanished on Eve of Surgery

When Brandon Connor was born with a tumour on his spine, doctors warned that any operation would carry considerable risks. Unsure as to which path to take, his parents spent the next two years anxiously watching for signs of deterioration. Every cold, every fever, every stomach cramp left them worried in case the tumour was spreading and was blasting their son's small body with its deadly cells. Eventually, after Brandon was sick for three weeks with a mysterious fever and abdominal pains, the Connors decided they could wait no longer and he was booked in for surgery in San Francisco. To everyone's amazement, one final pre-operation scan showed that the tumour had disappeared completely. Try as they might, the doctors could offer no explanation for this.

Doctors in Brandon's home city of Atlanta, Georgia, USA had first spotted the marble-sized tumour on an ultrasound when Kristin Connor was eight months pregnant. She had been sick throughout her pregnancy. "It had been such a long road, I couldn't believe there could possibly be something wrong with him," she said. "We were just devastated." They didn't know for certain what it was until Brandon was five weeks old when doctors diagnosed it as a neuroblastoma, one of the most aggressive childhood cancers. A neuroblastoma originates in neural cells and is often located near the adrenal glands, toward the small of the back. Less commonly, neuroblastomas develop in the sympathetic nerves of the chest or neck, or, occasionally, in the brain. About 80 per cent of cases develop during the first ten years of life, most commonly in the first four years. Neuroblastomas range from the relatively harmless to the highly malignant and for children who are not diagnosed until the disease has reached other organs, the two-year survival rate is less than 40 per cent. Neuroblastomas account for 15 per cent of all childhood cancer deaths.

With a tumour on the spine, there was a real risk that surgery to remove it might result in paralysis; on the other hand, to leave it unchecked could result in death. This placed the doctors treating Brandon in a dilemma. Eventually, because some neuroblastomas spontaneously regress before a child reaches one year of age, they decided that the best course of action was to monitor the tumour's progress by means of regular magnetic resonance imaging (MRI) scans. The Connors still did not know what to do for the best, and their hearts sank whenever Brandon went down with a bug or a stomach ache, which could have been symptoms of the cancer. And all the while scans showed that the tumour was still there.

As Kristin and Mike Connor began searching for more information about this little-known cancer, they came across neuroblastoma expert Dr Katherine Matthay, head of paediatric oncology at the University of California in San Francisco. She in turn consulted with a colleague, leading neurosurgeon Dr Nalin Gupta, who told the Connors that he thought he could remove the tumour without paralysing Brandon. But with the stakes so high, the family still remained undecided.

Then, in August 2003, with Brandon a few weeks short of his second birthday, they had a terrible scare. "He started running a fever," recalled Kristin. "At first it was 99, then 103. He stood up in the bathtub and cried for 45 minutes, 'Mommy, hurt, hurt!' The doctors thought it was full blown." Although aware that there was still a risk, they decided to go ahead with the surgery. So while their other son, five-year-old Ryan, stayed behind with his grandparents,

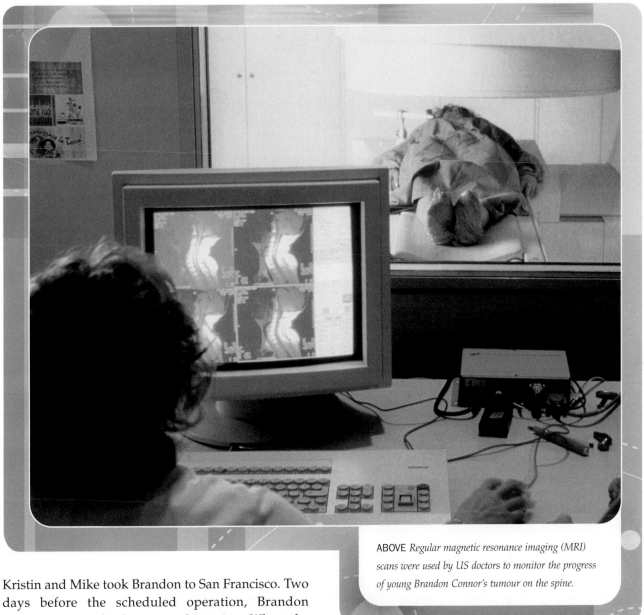

ABOVE *Regular magnetic resonance imaging (MRI) scans were used by US doctors to monitor the progress of young Brandon Connor's tumour on the spine.*

Kristin and Mike took Brandon to San Francisco. Two days before the scheduled operation, Brandon underwent one final scan on his spine. When the doctors looked at the MRI that evening, they could not believe what they saw. The tumour had gone; there was only fatty tissue.

Kristin Connor remembered: "Dr Gupta asked, 'Do you want the bad news or the good news first?' Of course I said the good news. He said, 'The good news is that the tumour has gone away. The bad news is that you came all the way to San Francisco for an MRI.' I was so overjoyed. Here it was, the twelfth hour, they said this couldn't happen. It's just a miracle."

The doctors were baffled by the tumour's sudden disappearance after two years but agreed that part of the problem is that so little is known about neuroblastoma. Brandon's local physician, Dr Bradley George, said: "Brandon's the only little guy with a peri-spinal mass that we've ever had that went away. Then again, we have not made the same progress with neuroblastoma that we have with other childhood cancers. We're frustrated: we don't know for sure how to treat them."

Dr Matthay added that the chances were good that the tumour would not return, but for the time being Mike Connor was just so grateful that his son had been granted an unexpected reprieve. "You don't want to ask why," he said. "You just take the gift and pass it on."

LEFT *A relieved Mike and Kristin Connor hug their two-year-old son Brandon after the boy's tumour mysteriously vanished on the eve of surgery.*

Woman Woke After 16 Years in a Catatonic State

For 16 years, Patricia White Bull remained unresponsive to the world – unable to speak, unable to swallow, and barely able to move her body. She had shown no indication of emerging from the catatonic state into which she had slipped during the birth of her fourth child. Then suddenly, on Christmas Eve 1999, while staff at the nursing home in Albuquerque, New Mexico, were routinely straightening the sheets on her bed, she sat up and snapped: "Don't do that!"

It was in 1983 that Ms White Bull, then 27, fell into a coma while her son Mark Junior was being delivered by Caesarean section. A blood clot lodged in her lung and caused her to stop breathing and, although doctors managed to resuscitate her, the oxygen to her brain was cut off for several minutes and she suffered brain damage. Her family were warned that there was little hope of recovery.

The White Bull family lived in a trailer in Edgewood, New Mexico, on five acres of open prairie near the Tijeras Canyon. Patricia's husband, Mark, was a computer programmer in Albuquerque, 35 miles away. Patricia was an activist with the group Indian Opportunity, but gave it up after having her first three children. She was always known to her friends and family as "Happi" because of her cheerful demeanour. On that warm June day in 1983, as Patricia left for the hospital to give birth to her fourth child, her parting words to her family were: "I'll see you tomorrow". That night, the three children – Jesse, Cindy and Floris – went to work with their father on the graveyard shift. When they stopped home on the way to the hospital the following morning, they saw a note attached to the door. Something had gone terribly wrong.

Mark Junior had been born healthy, but Patricia had not fared so well and was lying in the hospital bed with a network of tubes attached to her body. Her eyes were wide open, but she wouldn't blink. She was living, but she wasn't alive.

For the next three years Patricia White Bull's husband kept the children nearby, but after doctors told him that she would stay in a coma for the rest of her life, he moved them to a reservation in South Dakota so that his family could help with their upbringing. Losing hope, he also sought a divorce.

As the doctors had feared, there was no change in her condition over the next 16 years, but then an unconnected epidemic provided the miracle the family had been praying for. In December 1999 a cold virus was working its way through the Las Palomas Nursing and Rehabilitation Center in Albuquerque, where Patricia was a patient. Dr Elliot Marcus prescribed amantadine, a flu prevention drug, for those affected. The drug is also sometimes used to stimulate people with Parkinson's disease or brain injuries. A week or so later, Patricia White Bull unexpectedly returned to the world, and, although they could not be certain, doctors believed her sudden awakening was a direct result of the stimulant she had been given.

"You could just instantly tell there was a difference," recalled daughter Cindy, talking about the reunion she thought would never happen. "Her face was lit up." Patricia had plenty of catching up to do, not least a first meeting with the son she had given birth to all those years ago. "This is your baby," said a nurse, to which Patricia murmured: "Junior?" For the teenager, it was the first time he had actually heard his mother's voice.

After 16 years of profound silence, Patricia's speech was surprisingly clear although, perhaps understandably, she was doing more listening than talking at first. Her hands, which had been clenched tightly for so long, loosened and she was able to climb from her wheelchair and take a few shaky steps. On one of her first excursions, following the Christmas awakening, she was wheeled through a local shopping mall, where she pointed to a pair of running shoes and announced: "I'm going to run".

The amount of catching up she had to do was not confined to her family. The world at large had changed dramatically. After all, when she slipped into the coma, Ronald Reagan was president, the Soviet Union existed and the Internet did not. When her sister visited her at the nursing home and made a call on a mobile phone, Patricia's eyes widened in amazement at the tiny gadget.

After the initial euphoria surrounding her recovery had died down, practicality set in. For three hours a day she underwent therapy to teach her how to walk again, to talk, to brush her hair and wash her hands. She had good days when she talked a little and interacted with family members as well as times when she didn't really do anything. While doctors sought the right combination of stimulants to help Patricia White Bull stay alert, fearing that without drugs she might return to her vegetative state, her therapists warned that she would probably also need assistance for the rest of her life. For her family, however, after 16 years of torment, just having her around was enough.

RIGHT *After 16 years in a coma, Patricia White Bull regained the power of speech in 1999. Was her sudden awakening the result of being given a flu prevention drug?*

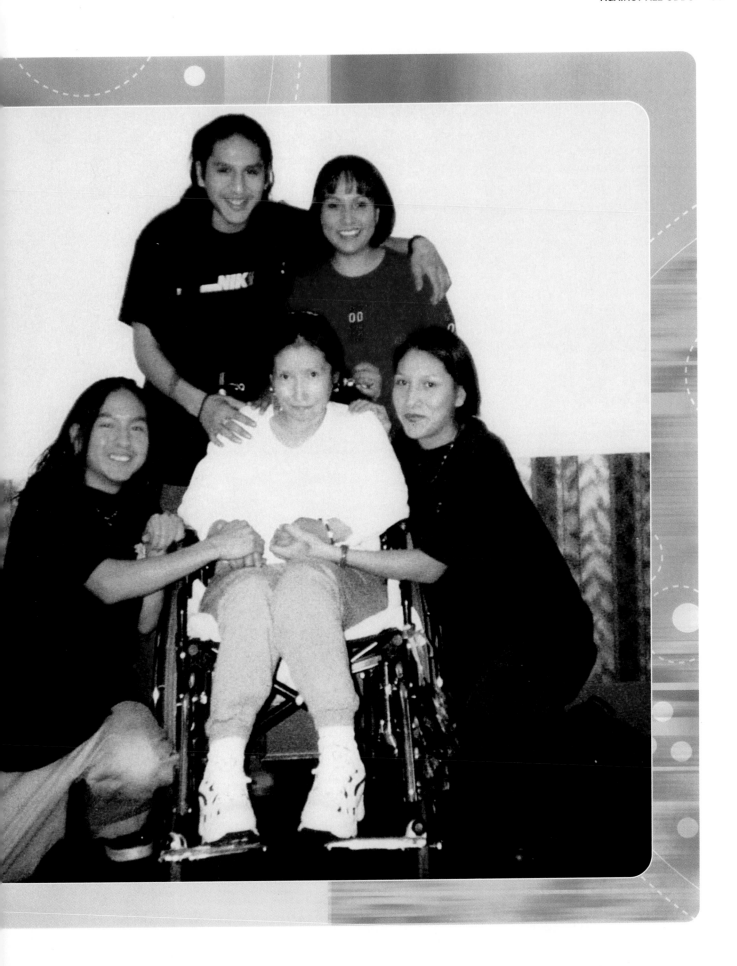

3 SPECIAL POWERS

The world has always been full of people who claim to possess powers that extend beyond normal human capabilities. In the past, such people often appeared in circuses or freak shows, demonstrating talents such as conducting electricity or bending metal. In truth, they were usually little more than magicians or illusionists, using cunning and chicanery to deceive an audience that was only too eager to believe. Others professed to be able to cure the sick with just a touch of the hand or a concentration of the mind, but the majority of these were eventually exposed as charlatans. However, from time to time someone makes a claim that, whilst extraordinary and beyond the comprehension of conventional medical science, is backed up by seemingly impressive evidence. In such instances even the experts are forced to sit up, take notice and at least investigate the possibility that there is some credibility to the stories. In actual fact, a number of conditions and phenomena that were once dismissed as flights of fancy have subsequently acquired recognized medical status. We still have a lot to learn about the human mind and body, so who is to say that some individuals do not genuinely possess healing hands or X-Ray vision?

Electric People

It is a riddle that has baffled great scientific minds for more than 150 years: how some people appear able to create an electric shock with their touch, act as human magnets by attracting metal objects to their bodies, or mysteriously make electrical appliances stop working.

A woman from Toronto, Canada, claims to possess the power to turn off streetlights in sequence; a woman from Tampa, Florida, says her sister disrupts electrical connections, especially the starter switches of cars, when she is upset; and a woman from California claims that she can blow out light bulbs and make computers crash.

Meanwhile a man from Ohio says he creates havoc with electrical equipment. In ten years he has got through five toasters, several cars and countless clock radios. "I was about to throw one [clock radio] out that had been broken for years when my son plugged it in and it worked fine. If I get near it, it doesn't work at all. I have had two electrical fires in two different houses. One was from a lamp that I turned on; the other was a light fixture in the kitchen ceiling. In both cases the wiring was inspected and found to be fine."

When a woman from California claimed to have put out thousands of lights on a huge Christmas display – including mechanical Santas, snowmen and reindeer – just by parking outside the decorated house, most readers would undoubtedly dismiss it as pure coincidence. But the same woman says she constantly disrupts her computer (even though technicians can find nothing wrong with it), receives large electric shocks from metal objects and often delivers electric shocks when shaking other people by the hand. So do these stories carry some perfectly innocent explanation or could there be something more mysterious afoot?

The concept of so-called "electric people" hit the headlines around the mid-nineteenth century with the case of "The Electric Girl", 14-year-old Angélique Cottin from Normandy, France. One evening in January 1846 Angélique and some other girls were weaving silk gloves on an oak frame when the frame suddenly began to twist and rock. It quickly became apparent that the frame only started to shake in Angélique's presence; when she was not around, it remained perfectly still. Over the ensuing days she

sparked many other curious incidents. Chairs twisted away from her when she tried to sit down, the power of the force being so great that a man was unable to hold down the furniture. A heavy table rose into the air when she touched it. If she attempted to sleep in a bed, it rocked violently, so that the only place she could rest was on a stone covered with cork. Whenever she went near objects, they moved away from her, even without any apparent physical contact. Similarly, people standing near her would receive electric shocks without her even touching them. When her power was at its most active, her heartbeat increased to 120 beats a minute.

Her worried parents asked a doctor to examine her. He noticed that her condition intensified when she was on bare earth, but that her powers were reduced when she was on a carpet or waxed cloth, or when she was tired. He also experienced a cold wind in her presence. Sometimes her powers ceased for a couple of days, only to start again without warning.

Eventually Angélique was referred to the Academy of Sciences in Paris, where a team of researchers, led by noted physicist François Arago, conducted a series of tests on "The Electric Girl". Arago observed that Angélique's force was at its most potent in the early evening, and that it seemed to emanate from her left wrist, inner left elbow and pelvis. Her body was affected by unpredictable movements and also by a sudden shaking, which was transferred to anyone who touched it. Arago noticed that the girl demonstrated a strange sensibility to magnets. When she approached the north pole of a magnet she experienced a powerful shock whereas the south pole produced no effect whatsoever. Even when a scientist changed the poles without her knowledge, she was always able to identify the north by the different sensations she felt. In addition she alternately attracted and repelled small objects in the same way as a magnet. Arago's conclusion was that Angélique possessed a form of electro-magnetism, possibly brought on by some kind of nervous malady. He wrote that her case demonstrated "…that, under peculiar conditions, the human organism gives forth a physical power which, without visible instruments, lifts heavy bodies, attracts or repels them, according to a law of polarity, overturns them, and produces the phenomena of sound."

After 12 weeks Angélique Cottin's special powers deserted her for good. However, her impoverished parents decided, against the advice of doctors, to exhibit her on the Paris stage as a paying attraction – she was persuaded to fake what may have originally been a genuinely mysterious phenomenon.

Another who made a living from her electricity was Annie May Abbott, alias "The Little Georgia Magnet". She toured the world demonstrating her apparent ability to raise a chair with a heavy man seated on it, just by touching it with her hand. How much of her talent was attributable to mere trickery remains unclear, but her audiences were impressed.

In France, in January 1869 a doctor at St Urbain delivered a baby who was so highly charged that he shocked anyone who touched him and luminous rays emanated from the infant's fingers. Alas the electric baby died at the age of nine months. In the following decade, Caroline Clare of Ontario, Canada, developed electrical powers soon after suffering a dramatic weight loss. Metal objects would leap into her hand and stick to her until someone pulled them off. She gave an electric shock to those she touched and in one experiment succeeded in passing the shock down a line of 20 people who were holding hands. Like Angélique Cottin, her powers lasted for just a few months and then vanished, never to return. Another case from that era was Louis Hamburger, a 16-year-old student from Maryland. When his fingertips were dry, he could apparently pick up heavy objects simply by touching them. Pins dangled from his open hand as if they were hanging from a magnet.

The *American Journal of Science* reported on the unusual events surrounding a lady from Orford, New Hampshire, who, while suffering from chronic rheumatism and neuralgia, suddenly began to discharge electricity. One evening while running her hand over her brother's face, sparks inexplicably shot out from her fingers. And when she stood on a thick carpet, sparks discharged around her hands. These attacks lasted for about six weeks and when they had finished, her ailments were miraculously cured.

The human nervous system does generate electricity. When we walk across a thick carpet, our body can build up around 10,000 volts, but because it can only develop a small electrical charge, the current that can be discharged is equally tiny. Somehow, "electric people" appear – on occasions at least – to be able to maximize their electrical potential.

A doctor who attended the New Hampshire woman remarked that her sparks were at their most spectacular during a heatwave and was convinced that the weather was partly responsible. Another theory is that human electricity is the after-effect of disease, and is governed by a person's health. However, until a thorough and convincing explanation is advanced, sceptics will continue to file "electric people" in the same drawer as "spoon benders" and "psychics".

ABOVE *Noted nineteenth-century physicist, François Arago, conducted a series of tests on Angélique Cottin, a French girl apparently blessed with extraordinary electro-magnetic powers.*

The Girl with X-Ray Vision

A 17-year-old girl who claims to have X-ray vision has astounded observers in her native Russia and also in Britain and Japan. Natalia Demkina from Saransk claims she is able to see inside human bodies – and can therefore discern the condition of a person's internal organs. She says she possesses dual vision and can switch from normal to "medical" vision simply by focusing on a person for about two minutes. However, she is apparently unable to see into her own body.

"Growing up, my daughter was just an ordinary child," says her mother, "although she was perhaps a bit more mature than other children her age. She started to talk when she was only six months old and learned to read when she was three years old." Then at the age of ten Natalia had her appendix taken out. Unfortunately, the doctors forgot to remove sanitary cotton packing from the girl's intestines, and she had to undergo a second operation. A month after the surgery she suddenly described her mother's internal organs in considerable detail even though she did not know the correct names for them. The family believe that their daughter's special powers were triggered by the bungled operation.

Her concerned mother took Natalia to a psychiatrist, and the girl proceeded to draw his stomach, complete with a previously diagnosed ulcer. As word of her talent spread, Natalia was subjected to rigorous tests at the main hospital in Saransk. In one test doctors showed her a girl who was very ill. Without knowing what was wrong with the patient, Natalia identified all of her illnesses. An ultrasound examination confirmed her diagnosis. On another occasion she was shown a lady who had been diagnosed with cancer. "I looked at her," says Natalia, "and did not notice anything like it, just a small cyst." Further examination revealed that Natalia was right. Although many doctors were naturally sceptical,

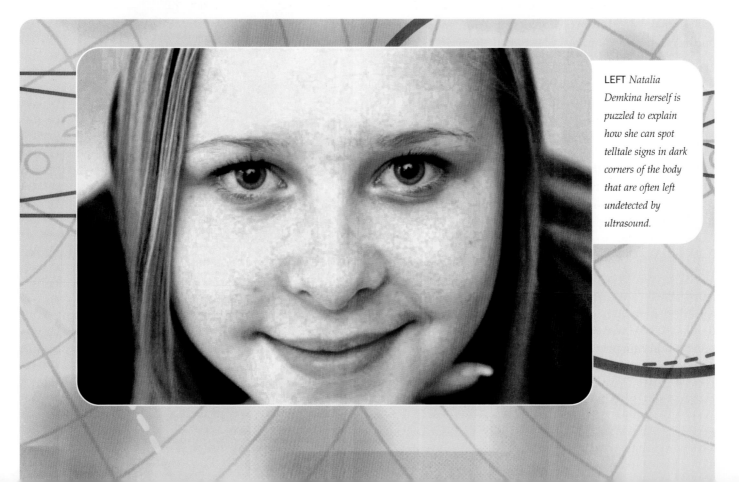

LEFT *Natalia Demkina herself is puzzled to explain how she can spot telltale signs in dark corners of the body that are often left undetected by ultrasound.*

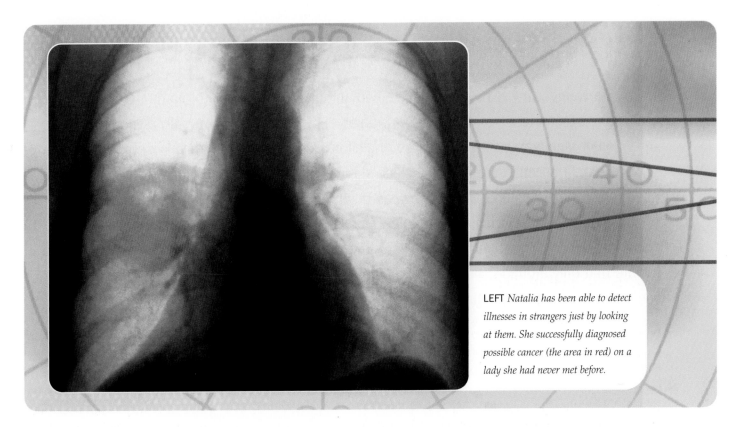

LEFT *Natalia has been able to detect illnesses in strangers just by looking at them. She successfully diagnosed possible cancer (the area in red) on a lady she had never met before.*

Irina Kachan, the senior consultant at the hospital, admitted: "The percentage of cases where she is right is very high".

In January 2004 she travelled to England to appear on the television programme *This Morning*. There, she correctly diagnosed the medical conditions of four strangers – a missing left kidney, damage to the spine, evidence of surgery on the spleen, and an old shoulder injury. The programme's resident doctor, Chris Steele, was impressed.

She appears capable of distinguishing even the tiniest sign in the darkest corners of the human body, areas that are often left undetected by regular ultrasound. "I see an entire human organism," she says. "It is difficult to explain how I determine specific illnesses. There are certain impulses that I feel from the damaged organs. My secondary vision works only in daytime and is asleep at night."

Russian scientists have so far been unable to come up with an explanation for the vivid and detailed accounts Natalia gives of what she sees when she looks inside the human body. Although a trip to the United States proved inconclusive when she was only able to determine the illnesses of four out of seven people, she impressed scientists in Japan by diagnosing someone's illness just by looking at his passport photo. From the tiny photograph, Natalia was immediately able to detect that the man had liver cancer. She also correctly diagnosed all seven patients who were brought before her for examination. Professor Yoshio Machi, from Tokyo University, who specialises in studying humans with seemingly special powers, said: "We did a whole range of tests, and the strangest thing was that we found she could use her abilities on photographs, even on tiny passport photos. She was able to look at them and apparently see what the problem was. She definitely has some kind of talent that we can't explain yet".

While sceptics remain unconvinced, back in Russia people have been begging her for consultations. She receives up to 20 phone calls a day and there are often queues of people outside her house. She never turns anyone away and never accepts any payment. She wants to be subjected to further experiments in the hope of finally coming up with some answers. "I have nothing to hide," she says. "Let them experiment with me. Perhaps they will be able to explain the nature of my secondary vision." In the meantime she is studying medicine at a Moscow academy. "Being able to use medical terminology, I will be able to state the final diagnosis more accurately. I have to know and understand what I see."

Foreign Accent Syndrome

In 1941 a young Norwegian woman sustained shrapnel injury to the brain during an air raid. At first she experienced severe language problems, but when she eventually overcame them, she was left facing an even more distressing situation – suddenly she could only talk in a strong German accent, as a result of which she was ostracised by her fellow Norwegians. Hers was the first documented case of what has become known as "foreign accent syndrome" (FAS) – a condition whereby patients who suffer a stroke, or some other form of head injury, occasionally find that they are no longer able to speak in their native accent. The condition is so rare that only around 20 cases have been reported since that wartime mystery in Norway. The majority of cases unwittingly adopt German, Swedish or Norwegian accents.

In 1990 Dr Dean Tippett, a neurophysiologist at the University of Maryland School of Medicine in Baltimore, USA reported the case of a 32-year-old local man who, a couple of days after suffering a stroke, inexplicably started speaking in a Scandinavian accent. Although nothing in the man's history suggested any knowledge of a foreign language, he suddenly sounded Nordic and was completely unfamiliar with English. He added extra, exaggerated vowel sounds as he spoke, the word "that" was pronounced as "dat", and his voice rose in pitch at the end of sentences. At first he apparently enjoyed his new accent, saying that he hoped it would attract women, but when his speech reverted to normal four months after the stroke, he was very happy to be talking like an American again.

Before November 1999, 47-year-old Englishwoman Wendy Hasnip spoke with a Yorkshire accent, but after suffering a minor stroke she could not stop talking in a French accent. Mrs Hasnip, whose only previous experience of the country was a weekend trip to Paris, said: "I was OK for a fortnight, then I began to stammer, which turned into an up and down voice. By the end of the week I was talking in a French accent. Yet I don't speak French, I didn't do O-level French, and I have no contact with France at all." Like the man from Baltimore, she found the phenomenon quite amusing. "I have been laughing my way through it from the beginning," she said. "There are much worse things than being left with a French accent."

Curiously, in the same month that Wendy Hasnip suffered unexpected consequences from her stroke, a similar case was being reported on the other side of the Atlantic. Tiffany Roberts, 57, had a stroke that left the right side of her body paralyzed. She was also unable to speak. After months of physical therapy, the paralysis had gone and she was able to speak, albeit with some difficulty. Over the next year her speech gradually improved until she was speaking with the same fluency that she had before the stroke. Instead of her familiar Indiana twang, she was now speaking in a British accent. Although she had never been to Britain, her voice had become a curious hybrid of Cockney and West Country accents and she had started using anglicisms such as "bloody" and "loo". Whereas once she spoke in deep tones, her voice was now a much higher pitch. She didn't even recognize her own voice, her friends and family didn't understand it, and strangers constantly asked her where she was from. One doctor told her she was not working hard enough to get her old voice back.

In a bid to regain her old pronunciation, she listened to a tape of herself before the paralysis. "For the first two years every day I would try and copy the tape and simulate the voice. I couldn't do it. So I would go to bed crying and wake up crying. At times I thought I was losing my mind. When people first started asking me where in England I was from and a family member asked why am I talking that way, that is when I became very conscious that a part of me had died during the stroke." After people in America accused her of lying when she said she was born and bred in Indiana, she began to avoid social contact and eventually developed agoraphobia, a fear

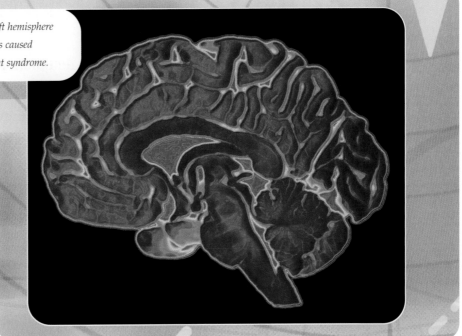

of open spaces. She became so desperate that she even thought about moving to England.

Then in 2003 – four years after her stroke – Mrs Roberts received an e-mail from a friend containing a *New York Times* article on language tests conducted by Dr Jennifer Gurd, a professor at Oxford University who had been researching foreign accent syndrome for 15 years. Whereas a number of doctors had dismissed the problem as a psychiatric disorder, Dr Gurd and her team of scientists were convinced that it was a physical condition and had made a significant breakthrough in the investigation of FAS.

Puzzled as to why only a tiny proportion of stroke victims developed FAS, they found that patients with the syndrome shared certain characteristics – namely small areas of damage to parts of the left hemisphere of the brain, where language processing occurs. It therefore followed that damage to those areas could result in altered pitch, lengthened syllables or mispronounced sounds, causing speech to sound accented. There is no need for patients to have been previously exposed to their new adopted accent because they haven't really picked it up; it is simply that the injury to their brain has caused their speech patterns to change and to make it sound as if they have acquired a foreign accent. The precise area of damage coupled with the severity of the trauma may also determine whether the condition lasts for a few weeks or several years.

Dr Gurd said: "What we find very interesting about the changes in accent is that they may be indicating that, in the human mind, there is a separate, independent module for accents and human language. The way we speak is an important part of our personality and influences the way people interact with us. It is understandably quite traumatic for patients to find that their accent has changed."

For Mrs Roberts it came as a relief to learn that her problem was neurological rather than psychological. She contacted Dr Jack Ryalls of the University of Central Florida, an expert on neurologically based speech and language disorders, who carried out a series of tests. He checked her ability to change word stress patterns and accentuation as well as making her repeat words stressing the wrong syllable. The tests enabled him to analyse her ability to use the appropriate syllable emphasis and accentuation, which differs in British and American English. He wanted to determine if she was applying the wrong syllable as part of her accent. He found that Mrs Roberts had developed unusual methods of coping with her new voice. Whenever someone asked her where in England she was from, she would counter: "Where do you think I'm from?" And whatever city was named, Mrs Roberts would say the person was right. "In some ways," said Dr Ryalls, "her response showed she was beginning to accept the accent. It was an ingenious coping mechanism, but it also reveals that she had begun to resign herself to the change in her speech."

Tiffany Roberts says that she wants people to know more about FAS in the hope that other sufferers can be spared the alienation that she has experienced. "If I can bring notice to this condition, especially within the medical community, doctors may be able to help others who find themselves in my situation."

Healing Hands

Psychic healing has been practised for centuries, and there has never been a shortage of people willing to put themselves forward as having the magic touch. Unsurprisingly, conventional doctors have tended to dismiss such individuals as frauds and charlatans. In many cases they have probably been right, but equally, as we shall see with the placebo effect, the cure for certain illnesses and ailments may rest in the mind as much as in the body. If the exclusively physical methods of the doctor or surgeon sometimes do not have the desired effect, it is perhaps to be expected. After all, the human body is such a complex creation and there is still a great deal to learn about the workings of the brain and its relationship with our emotions. So maybe healers sometimes succeed where standard medicine has failed simply because they are trying a different approach, by appealing to our mind rather than our body. If we have complete faith in being cured – healers know exactly what to say to encourage that sense of belief – then there could be some tangible benefits to what they do.

Certainly those who visited Harry Edwards thought so. In his heyday in the 1950s, the celebrated British healer could fill the 8,000-capacity Royal Albert Hall in London with people eager to witness – and benefit from – his seemingly remarkable powers. There was no mystic aura to Edwards: he usually just rolled up his sleeves and placed his hands on the affected part of the patient's body. A printer by trade, he only took up his new career in his 40s after being told by a number of mediums that he had a latent ability as a healer. One of his first patients was a young girl who was suffering from tuberculosis of the lung. When Edwards put his hands on the girl's head, he claimed that his entire body was suddenly filled with energy, which rushed down his arms, out of his hands and in to the patient. As the strange sensation eased, he heard himself telling the girl's mother that her daughter would be out of bed in three days. Not only did that prophecy come true, but at her next medical examination the girl was pronounced completely cured.

While the concept of "hands-on" healing is reasonably accessible, that of distant healing – where the healer does not even meet the patient – is much harder to fathom. Arguably the most famous exponent of distant healing was the American Edgar Cayce. Born into a farming family near Hopkinsville, Kentucky, in 1877, Cayce worked as a salesman until he was 21, but had to give up his job when he lost his voice. Having suffered repeated bouts of laryngitis, he consulted a hypnotist, Al Layne, in the hope of finding a cure. Lane apparently put him in a trance and asked him to identify the cause of his illness plus a likely remedy. Cayce suggested that his voice loss was due to psychological paralysis and that it could

ABOVE *British faith healer Harry Edwards treating a four-year-old boy, who was unable to walk due to a spinal injury, at London's Royal Festival Hall in 1951.*

be corrected by increasing the blood flow to the voice box. As his face became flushed with blood, he awoke and discovered that he had regained his voice.

Now convinced that healing others was his destiny, Cayce set about advertising his services and received letters from far and wide. Armed with no more than the name and address of the patient, he would put

himself into a trance. His assistant (usually Cayce's wife) would read out the name and address and tell him: "You will go over this body carefully, noting its condition and any parts that are ailing. You will give the cause of such ailments and suggest treatments to bring about a cure". Cayce then commenced the reading. Having located the subject – sometimes

LEFT *American Edgar Cayce, a former salesman who became a celebrated faith healer despite rarely ever meeting his patients.*

naming streets along the way – he proceeded to examine the correspondent's body in his trance-like state, describing how the various organs were functioning, identifying any areas of concern and suggesting reasons for the problem. Like Natalia Demkina (p.60), he claimed to be able to see every nerve, gland, blood vessel and organ inside the patient's body. Cayce believed that every single cell in the body was individually conscious and maintained that the patient's cells actually communicated their condition to his entranced mind. Finally, he would prescribe a cure. This might range from orthodox drugs or surgery to massage, obscure herbal remedies and the relatively untried practices of hydrotherapy and osteopathy. Since most of his patients had only written to him as a last resort, after all conventional means of treatment had failed, he also emphasized that their mental state could greatly aid their recovery. The medical profession of the day ridiculed his

revolutionary ideas that diet and stress could cause illness, but now we know that he talked more sense than he was given credit for. Unfortunately, he did not help himself by also claiming that he was a reincarnation of an angel who inhabited Earth before Adam and Eve and that in yet another life he had been an inhabitant of Atlantis.

Ironically, Edgar Cayce's readings adversely affected his own health. He gave over 9,000 physical readings in his lifetime, but the energy expended left him exhausted. Despite being warned that giving more than two readings a day would kill him, by the time the USA had joined in the Second World War he was averaging up to six a day. In August 1944 he collapsed through exhaustion and five months later he was dead, leaving behind thousands of satisfied customers.

One of the best-known modern healers is English psychic Matthew Manning who, simply by touch and concentrating his mind, claims to be able to alter the electrical resistance of the human skin and to speed up the death of certain types of cancer cells. On a tour of West Germany in 1981, doctors who examined Manning's patients before and after his healing sessions reported an immediate 95 per cent improvement. In Freiburg, Manning met the wife of an independent medical consultant, Dr Otto Ripprich. After sustaining crippling nerve and muscle damage in an accident, Frau Ripprich had been unable to straighten her right arm for several months, but, to the amazement of her husband, she was able to straighten her arm fully after just five minutes of treatment from Manning. Other patients have reported that Manning has helped cure them of all

manner of conditions from tumours and cancer of the vocal chords to poor eyesight and hay fever.

In certain parts of the world – notably Brazil and the Philippines – the work of psychic healers is complemented by psychic surgeons who, using neither anaesthetics nor conventional surgical instruments, claim to be able to perform most of the feats of modern medicine with their bare hands. The leading Filipino psychic surgeon is Alex Orbito who, while travelling the world, has treated over a million people, ranging from actress Shirley MacLaine to members of the Saudi royal family. One of his earliest patients was a woman with abdominal pains. According to Orbito himself, he was praying with his hands over the woman's stomach, gently kneading and trying to transmit the energies that would expel the negativity he felt beneath his fingers. Suddenly, there was a faint sound as his fingertips felt a warm and wet sensation. Opening his eyes, he was horrified to see that his fingers had penetrated the woman's abdomen, creating a small puddle of blood. Fearing that the woman was dead, he promised to be more careful in future, but as he withdrew his fingers and closed his hands, he saw that the opening had vanished and all that remained was a thin red line with a few drops of blood, which were easily removed with a handkerchief. The woman was fine and amazingly she had felt nothing.

Investigative journalists have exposed some Filipino "surgeons" as frauds, but with others the case remains as inconclusive as the whole field of psychic medicine. Until someone can produce concrete evidence one way or the other, those who want to believe, will and those who don't, won't.

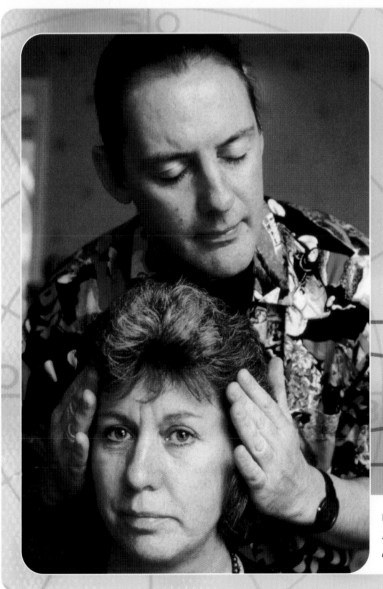

LEFT *Among the best-known modern healers is Matthew Manning who, by touch and concentration, claims to be able to speed up the death of certain types of cancer cells.*

Human Antennae

Among the most common urban legends is that we can sometimes hear radio broadcasts through our teeth. Although such stories are invariably dismissed as a figment of the imagination, they have been doing the rounds since the days of Marconi and still resurface from time to time. In fact the American Dental Association says it receives inquiries on the subject virtually every month.

A man from Chicago said that after losing a tooth as a boy, around 1960, he was fitted with a cap that was attached to the tooth stump by a wire. Thereafter he apparently began hearing music in his head, usually while he was outdoors. He described the music as soft and distinct, but he was unable to identify the radio station. A year or two later a new dentist put in a cap without a wire and the tunes stopped. A fellow American recounted a similar experience back in 1947 when she was travelling by train from her home in Cleveland to college in Rhode Island. She said she picked up a radio station in her head for about ten

ABOVE *Can some people really be human antennae or are the apparent sounds of radio waves nothing more than a chemical reaction in the mouth?*

minutes and recalled hearing commercials and an announcer's voice. She had silver tooth fillings, but couldn't remember if one had been put in just before her musical interlude.

Most famously, comedienne Lucille Ball once claimed that, shortly after having several temporary lead fillings installed in her teeth, she was driving through California one evening in 1942 when she suddenly started to hear music. "I reached down to turn the radio off," she wrote, "and it wasn't on. The music kept getting louder and louder, and then I realised it was coming from my mouth. I even recognised the tune. My mouth was humming and thumping with the drumbeat, and I thought I was losing my mind. I thought, what the hell is this? Then it started to subside." When she supposedly recounted the story to actor Buster Keaton at the studio the next day, he laughingly told her that she was picking up radio broadcasts through her fillings, and that the same thing had happened to a friend of his. Of course, the entire story may be Hollywood embellishment, but back in the 1930s and 1940s, when powerful AM transmitters were installed around the United States, a number of local residents reported music coming from fence wire, bathtubs and tooth fillings. So is it all just folklore or might there be some scientific grain of truth in the matter?

Some scientists say that, given the right conditions, a person's mouth can act as a receiver circuit. At its most basic, a receiver circuit consists of only three elements: an antenna, which picks up an electromagnetic radio signal; a detector, which is an electrical component that converts the radio wave to an audio signal that the human ear can pick up; and a transducer, which is anything that acts like a speaker. They claim that, on rare occasions, the human mouth could replicate this set-up. The electrical conductivity of the human body can act as an antenna. A metallic filling in a tooth, reacting with saliva, can act as a semiconductor to detect the audio signal. And the speaker could be anything that vibrates within the mouth sufficiently to produce noise, such as bridgework or a loose filling.

Others dismiss the notion out of hand, suggesting that what sounds like radio waves is really just a chemical reaction caused by a strange interaction between the fillings in the mouth and the acid in saliva. In other words, it's just wishful thinking.

Either way, although the occasional report still filters through, the heyday of hearing music through teeth seems to have been at least 40 years ago. Perhaps it had something to do with the old-fashioned radios or the type of metal used in fillings? We may never know.

ABOVE *In 1947 an American woman with silver tooth fillings said she picked up a radio station in her head for about ten minutes.*

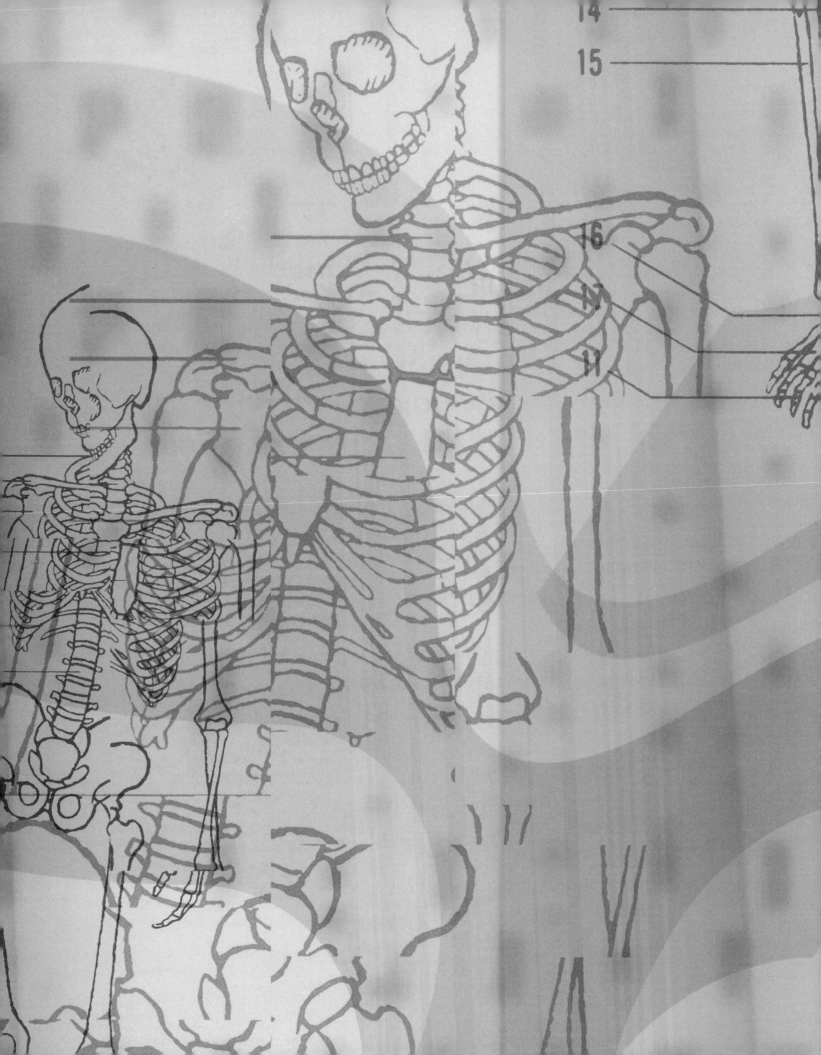

14

15

16

17

11

THE AMAZING BODY

4

On average a pair of human feet lose half a pint of water in perspiration each day. Relative to size, the strongest muscle in the human body is the tongue. The average person passes about 8,800gal (33,310 l) of urine in a lifetime – enough to fill 500 baths. There is enough fat in the human body to make seven bars of soap. And a sneeze travels at speeds of over 100mph (161kmph). The human body houses hundreds of mind-boggling statistics yet for the most part we take it for granted. Sneezing, blinking, coughing, stretching, sleeping, yawning, walking – we rarely give them a moment's thought unless we are temporarily or permanently deprived of that particular function. Otherwise they just occur naturally and we don't think to question how or why they operate. Yet everything exists for a purpose – even eyebrows and toenails – and just as some eminent physicians are pushing back the boundaries of science in search of life-saving remedies, others are devoting their time to exploring the minutiae of our everyday existence in the hope of solving long-standing medical puzzles. While they attempt to unravel the mysteries of sleepwalking, hiccupping and tickling, there are always odd cases waiting to surprise even the most seasoned students of the human body – few more amazing than that of the boy who turned out to be pregnant.

The Boy who was Pregnant

For the first seven years of his life Alamjan Nematilaev had a distended stomach. His parents thought he had rickets, a common childhood disease in his native Kazakhstan. Teased at school because he looked pregnant, Alamjan could only watch helplessly as his stomach continued to grow. One day in 2003, after a gym lesson, he complained that he could feel something moving inside him. Alarmed by the bulge, which surgeons later described as resembling a woman in her sixth month of pregnancy, the school doctor examined him and insisted that the boy go straight to hospital. There, doctors scanned his belly and identified what they believed to be a huge cyst. They operated the next day and discovered a large, rounded mass pushing against Alamjan's stomach and lungs. After removing it in a delicate operation, they cut through the sac that surrounded it and saw dark hair, arms, fingers and nails, legs, toes, genitals, a head, and an approximation of a face, but they were still unsure as to exactly what it was.

LEFT *Alamjan Nematilaev's parents attributed his distended stomach to rickets, but it turned out that he was carrying the living foetus of his unborn twin brother.*

Instances have been reported where adults have had cysts removed that turn out to contain hair and teeth. Dr Virginia Baldwin, a Vancouver-based paediatric pathologist, diagnosed Alamjan's case, however, as being one of foetus in foetu. This is a rare condition where one healthy twin foetus grows around the other at an early stage of development, the incomplete foetus liveing off the healthy foetus as a parasite. The foetus, which measured some 8in (20cm) long, had remained alive inside Alamjan's stomach by attaching itself to his blood vessels.

In the past 200 years only 70 cases of foetus in foetu have been identified, but Dr Baldwin believes the condition is more common than these figures suggest.

"When twins are conceived there can be competition for resources and only one may survive. Depending on the anatomy and physiology of the placenta shared by the twins, a situation may develop that favours one or the other. If there is an imbalance in the blood flow, you may have a hazardous situation. When you talk to women who are very sensitive to the messages their bodies are giving them, they will say that they knew something was wrong, but didn't know what.

There may be no sign. If an abnormal twin developed in the very early stages and didn't survive, it would often disappear without a trace."

Doctors are uncertain as to what causes the condition. One theory is that it is simply one of the risks of the embryonic development of twins. Twins arise either because two separate eggs are fertilized by separate sperm or because a single fertilized egg divides into two. The former are known as fraternal twins, the latter identical twins. One of the earliest places that germ cells, which eventually produce the genitals, develop is the yolk sac attached to the embryo. In rare instances the two yolk sacs of identical twins can end up connecting. If one baby's heart develops before the other, the connection will make blood circulate from the healthy baby into the yolk sac and backward into the arteries of the less developed baby. That could prevent the second baby's heart from developing. As the embryo grows further, the yolk sac is normally drawn back into the foetus. In the case of a foetus in foetu, the healthy baby would draw in what is left of its twin along with its yolk sac. If the internal twin, which develops as a parasite of the healthy baby, gets a lot of blood it can, as in the case of Alamjan's brother, remain alive and acquire recognizable features, such as legs and fingers.

"It was remarkable," said Dr Valentina Vostrikova, leader of the medical team that operated on Alamjan. "We couldn't believe our eyes when we scanned him. We could see the clear shape of a baby inside him. And it wasn't a small baby. For almost seven years it lived like a parasite inside the boy's body. The embryo was recognizably male and lay in such a way that he sustained himself from his brother. Technically the baby was still alive, yet was not sustainable when separated from his brother. We had never heard of a case like it. Thank God the school doctor insisted on taking him to hospital. If this had gone on, we would not have been able to rescue him."

Alamjan recovered from his ordeal and his parents, not wishing to subject him to psychological trauma, kept the story of his "pregnancy" a secret from him and simply told him that he had become ill from eating unwashed fruit. His mother Gulnara admitted that she had told him off for making up stories when he had complained of something moving inside him. "I hadn't listened properly and told him to be quiet," she sobbed. "I almost fainted when doctors broke the news. I was stunned. It's just not what you expect to hear. We knew he was a bit overweight, but pregnant...?"

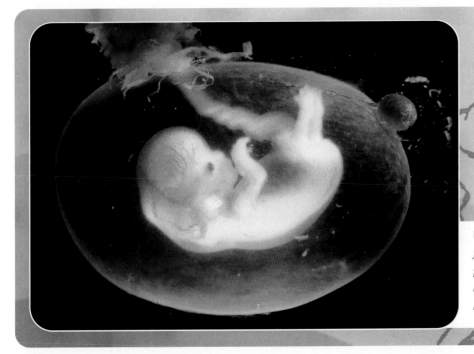

LEFT *Only 70 cases of foetus in foetu have been identified over the past 200 years. It is thought to be one of the risks of the embryonic development of twins.*

Why do we Hiccup?

The reason why humans hiccup has baffled scientists for centuries, not least because it does not appear to serve any useful purpose. Far from being of any benefit, hiccups are a nuisance, particularly if, like Charles Osborne of Anthon, Iowa, you have had them for 68 years! The attack started in 1922 while he was attempting to weigh a hog before slaughtering it and continued unabated until 1990 – an estimated 430 million hiccups later. Ironically, he died the following year.

Fortunately most bouts of hiccups are considerably less extreme and can be cured by a variety of popular remedies (drinking water, holding your breath, being slapped on the back) in a matter of minutes. A hiccup is an irritation of the diaphragm that causes a spasm. For the most part, the diaphragm works perfectly. It pulls down when we inhale to help pull air into the lungs, and it pushes up when we exhale to help push air out of the lungs. But sometimes the diaphragm contracts involuntarily because the nerves that control it have become irritated. The most common cause is eating or drinking something too quickly, the irritation resulting from endeavouring to breathe and eat at the same time. When this happens and air is inhaled, the space between the vocal cords at the back of the throat (the glottis) snaps shut with a clicking sound. That sound is what we hear when we hiccup.

But although there is no mystery as to what causes hiccups, their precise purpose has eluded even the finest medical minds through the years. In an attempt to find an explanation, scientists turned to the first stages of life. Ultrasound scans reveal that two-month-old babies hiccup in the womb, even before any breathing movements appear. One theory is that the contractions prepare the unborn baby's respiratory muscles for breathing after birth; another is that they prevent amniotic fluid entering the lungs. However, neither of these theories explains all of the features of hiccups. For example, if their purpose were to prevent liquid getting into the lungs, it would make more sense for them to involve a cough-like breath out as opposed to a breath inward.

Then, in February 2003, French scientists came up with a new theory. A group of researchers led by Christian Straus at Pitie-Saltpetriere Hospital in Paris suggested that the reason we hiccup could be linked to evolution and the fact that our ancestors lived in the sea. The Paris team pointed out that there is one group of animals for which combining closure of the glottis and contraction of the breathing in muscles does serve a definite purpose – primitive air breathers such as lungfish and frogs that still possess gills. These creatures push water across their gills by squeezing their mouth cavity while closing the glottis to stop water entering the lungs. Perhaps, argued Straus, the brain circuitry controlling gill ventilation in these early ancestors has persisted into modern mammals, including humans.

Epiglottis

Glottis

Larynx

ABOVE *When an irritation of the diaphragm causes a spasm and air is inhaled, the space between the vocal cords at the back of the throat snaps shut with the familiar hiccup sound.*

The researchers illustrated a number of similarities between hiccuping and gill ventilation in creatures such as tadpoles. Both are inhibited when the lungs are inflated and by high carbon dioxide levels in air or water. As for why humans still hiccup, 370 million years after our ancestors began hauling themselves on to land, Straus believed the circuitry that controlled the movements of the gills and glottis was conserved during evolution because it formed a building block for more complex motor patterns, such as suckling. The sequence of movements during suckling is similar to hiccupping, with the glottis closing to prevent milk entering the lungs. "Hiccups may be the price to pay to keep this useful pattern generator," said Straus.

Meanwhile, a 50-year-old man from Texas had non-stop hiccups for a year following a stroke. Hiccups are sometimes linked to irritation of the nerves that extend from the neck to the chest, and in the case of Shane Shafer it is thought that the problem was connected to an irregularity with the vagus nerve, brought on by the stroke. Surgeons suspected that the stroke had triggered an abnormal connection between cells in the brain associated with the vagus nerve, and other cell groups associated with the phrenic nerve that supplies the diaphragm. The persistent hiccups caused him such distress that he needed ten painkiller injections a day to provide a degree of relief, the only alternative being to make himself vomit. But in 2004 he underwent pioneering surgery at Louisiana State University that involved the fitting of a device called a Vagus Nerve Stimulator, which controls the stimulation of the nerve. It uses electrical impulses delivered by a generator implanted in the patient's chest to two tiny leads wrapped around the nerve in the neck. As soon as the implant was activated, Shafer's hiccups stopped.

ABOVE *The reason why we hiccup could date back to primitive air breathers such as frogs, which still possess gills.*

BELOW *Hiccups can occasionally be linked to a stroke. In extreme cases bouts of hiccups may last for a year.*

Why are Yawns Contagious?

Why is it that when we see someone yawn, we instinctively yawn ourselves? Although yawning is traditionally associated with tiredness or boredom, neither of these conditions appears necessary for us to yawn in response to another person. After all, we can be perfectly alert and awake, but we just can't help ourselves if we see someone in the same room (or even on television) start to yawn. In extreme cases people have been known to yawn just by reading about or thinking about yawning. So what do we know about this enduring medical riddle?

Dr Robert Provine, professor of psychology at the University of Maryland and one of the world's foremost authorities on yawning, has studied the subject at length. He has found that yawning serves a purpose in opening the Eustachian tubes, which run from the ears to the throat, and in adjusting the air pressure in the middle ear. It also plays a key therapeutic role in preventing respiratory complications following an operation. Interestingly it has been stated that schizophrenics rarely yawn, unless they are suffering from brain damage, and that those with critical physical illnesses do not yawn until they are on the way to recovery.

Provine also found that, although looking at a picture of a yawning mouth does not make us yawn, looking at the eyes of a yawner does. He was able to dispel the popular assumption that we yawn in response to low oxygen or high carbon dioxide levels in the blood or brain through tests that revealed that subjects do not yawn more when breathing air with enhanced levels of carbon dioxide, nor do they yawn less when breathing pure oxygen. Having observed that Olympic athletes yawn before their events, he was also able to refute the suggestion that yawning is wholly linked to tiredness or boredom. Instead, he concluded that yawns help our bodies change state from activity to inactivity and vice-versa, which is why we tend to yawn immediately before and after sleep. The yawn is a way of either relaxing or stimulating the brain, depending on the occasion. As for contagious yawning, Provine believes that it is leftover from the time when we were tribal creatures and that yawning together helped to synchronize our tribes. When we were tired, we started to yawn and that way everyone realized that it was time to go to sleep.

What adds to the mystery about contagious yawning is that it seems largely unconscious. It appears to be an automatic response, triggered by our brains. We see someone yawn, develop a sudden urge to mimic it and do it; all without even thinking about it. On other occasions we may be aware that we are doing it, but don't really know why. Research has shown that when adult humans are shown videos of yawns, up to 55 per cent also begin yawning themselves. Indeed, it does seem that just being around someone who is yawning is not enough; in the majority of cases we actually have to see a yawn to start yawning ourselves. To underline this point, it has been found that those subjects that perform contagious yawning recognize images of their own faces and are better at deducing what other people are thinking from their facial expressions. Brain imaging studies have also shown that people watching others yawning have more activity in parts of the brain associated with self-information processing.

Gordon Gallup, an evolutionary psychologist at the State University of New York, told *New Scientist*: "Our data suggest that contagious yawning is a by-product of the ability to conceive of yourself and to use your experience to make inferences about comparable experiences and mental states in others." This theory is backed up by the behaviour of small children. Human infants do not start to recognize themselves in mirrors until they are two years old and they also do not yawn contagiously. Similarly people with schizophrenia have problems with self-awareness and show little or no evidence of contagious yawning.

Recently, a team from the Helsinki University of Technology in Finland carried out further investigations into the matter. While volunteers

ABOVE *In a bid to uncover the mysteries of the yawn, experts believe that it plays an important role in preventing respiratory complications after an operation.*

watched videos of actors yawning or making other mouth movements, their brains were scanned using functional Magnetic Resonance Imaging, a system that shows the amount of activity going on in various areas of the brain, based on the amount of oxygen being used up. The volunteers were later asked how strongly they had been tempted to yawn while viewing the pictures. The study confirmed that contagious yawning is largely unconscious. Wherever it might affect the brain, it bypasses the known brain circuitry for consciously analyzing and mimicking other people's actions. This circuitry is called the "mirror neuron system" because it contains special neurons that become active both when their owner does something and when he or she consciously imitates someone else's actions. However, the Finnish researchers found that these brain cells were no more active during contagious yawning than during any non-contagious facial movements. Accordingly, they concluded that brain activity "…associated with viewing another person yawn seems to circumvent the essential parts of the MNS (mirror neuron system), in line with the nature of contagious yawns as automatically released behavioural acts, rather than truly imitated motor patterns that would require detailed action understanding."

The Helsinki team also noted an apparent deactivation of a second area of the brain, the left periamygdalar region, during contagious yawning. The periamygdalar region has been linked to the unconscious analysis of emotional facial expressions. The more strongly a subject reported wanting to yawn in response to another person's yawn, the stronger the deactivation in that area. This finding represents the first known neurophysiological signature of perceived yawn contagiousness, although as yet no definite conclusions have been drawn from this discovery. Indeed, apart from the fact that it is somehow connected to the brain, the reason for contagious yawning remains as much a mystery as ever.

In 2004 a research team led by James Anderson at the University of Stirling in Scotland discovered that yawning isn't only catching among humans: chimpanzees do it too, so perhaps the chimps will be able to shed some light on the subject?

ABOVE *Yawning is not only contagious among humans. Just seeing another person (or lion) yawn seems to be enough to set off imitational behaviour.*

What are Eyebrows for?

Neanderthal man was renowned for having hair all over his body, but as he evolved further from the apes he started to lose most of it, particularly from his face. One area that has mysteriously remained hairy on human faces is the brow area above the eyes. This is not the only puzzle connected to the twin caterpillars perched above our eyes because one of the most enduring medical questions is: what exactly are eyebrows for?

At first glance they appear to serve little purpose, save as a decorative space for modern teenagers who use them as somewhere to attach rings and studs. We can't see through our eyebrows, we can't hear through them, we can't smell through them, so what is their purpose?

In fact, far from having no practical use, scientific studies have unearthed no fewer than four plausible reasons why we have eyebrows. The first is linked to one of the few things we can actually do with them – move them. Eyebrows play an important role in non-verbal communication. As with all of the facial muscles, those around the eye constantly react, subconsciously, to our thoughts and emotions. A raised eyebrow immediately expresses surprise and a furrowed brow shows concern – in some instances of communication, therefore, eyebrows may be even more effective than words.

A second, and probably the most popular, theory is that eyebrows divert rain or sweat down our cheeks, thereby keeping our eyes dry and clear to see. This may also explain why eyebrows have remained hairy – unless, of course, you pluck them. The hairs, plus the arched shape of the brow, help to guide fluid sideways and away from the eyes. Additionally, the hairy eyebrows soak up stray beads of salty sweat. This was obviously an essential feature for early man to retain as it is difficult to outrun a predator if you can't see where you are going.

A third possible reason for the existence of eyebrows lies with the fact that, in order for the upper eyelid to work, the muscles that form the lid have to pull themselves up against something. That something happens to be the eyebrow, with the eyelid muscles attaching themselves to the structure of the brow like a curtain to a rail.

A fourth explanation is that the bony protuberance of the eyebrow, along with other facial bones, serves to protect the soft part of the eyeball. Without the bones that surround and guard the eye, any blow to the face would dangerously damage the eyes.

One other mystery about eyebrows: why is it that when some people pluck their eyebrows, they sneeze? It could be that in pulling the hair, we disturb tiny dust particles which in turn aggravate our nasal passage, but it also seems that eyebrow plucking excites a branch of the nerve that supplies our nasal passage. Any irritation of the nasal passage excites our trigeminal nerve, the branches of which supply sensation to the face, including the upper eyelids and nose. The impulses from the nervous system travel to a set of neurons located in the brainstem, which have collectively been termed the "sneezing centre". Doctors speculate that even though the impulses aren't coming from the nose, the stimulation generated by eyebrow plucking makes the entire nerve more sensitive, and enables sufficient impulses to reach the "sneezing centre" to produce a sneeze.

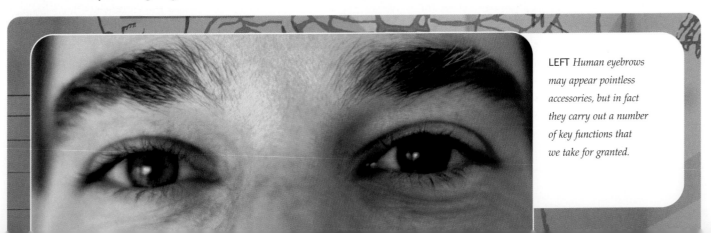

LEFT *Human eyebrows may appear pointless accessories, but in fact they carry out a number of key functions that we take for granted.*

Can Pain Predict Rain?

The alarming rise in the number of cases of skin cancer caused by exposure to the Sun's rays has made all of us more aware of the connection between climate and health. At the other end of the thermometer, frostbite is an obvious consequence of exposure to extreme cold, while hay fever always seems worse on windy days. While the dangers of skin cancer have only been highlighted recently, the link between some ailments and the weather dates back to at least Hippocrates in the fourth century BC, with countless old wives' tales reporting a connection between rain and pain. Most of us know someone who claims to be "weather sensitive", usually an elderly aunt who will stare out of the window on a sunny day, rub her arthritic shoulder and declare solemnly: "It's going to rain later".

So is there any scientific link between pain in an arthritic joint and wet weather? The evidence is inconclusive. One of the first studies into the subject was conducted by a scientist named Edstrom in 1948. He found that people with rheumatoid arthritis felt better when kept in a constant environment of warmth and dryness. Then, in 1961, Dr Joseph Hollander of the University of Pennsylvania Medical School put 12 people (eight with rheumatoid arthritis, four with osteoarthritis) in a special "weather chamber", in which the temperature, barometric pressure and humidity could be adjusted. Eight of the dozen had stated in advance that they were weather sensitive, and seven of those eight suffered a worsening of their symptoms with increased humidity and decreased barometric pressure.

A drop in barometric pressure often precedes a storm and the theory is that a decrease in the air pressure can cause the tissues around the joints to swell, causing arthritic pain. This may be the result of cell permeability. Blood vessel walls are invariably more permeable among arthritis sufferers so that more blood fluid is forced into the tissues. Blood is always under higher pressure than the surrounding body tissues and the movement would be greatest when the pressure of the environment outside the body is lowest – as it is when the barometer falls. If joints are already sore and swollen, the added fluid could trigger the additional pain. To support this idea, experiments have been conducted using a balloon in a barometric chamber as a simulator. When the pressure outside drops, the air in the balloon expands.

If this were to be replicated around an arthritic joint, the increased swelling could irritate the nerves, leading to pain. It could be that the sensitivity of the nerves is so highly tuned to barometric pressure that they can respond to even minor changes.

This explanation sounds highly plausible, but at the moment it remains just a theory as nothing has been proved, scientifically, to back up the claim. This is partly because any swelling in the human joints that is caused by a drop in barometric pressure would be so small that it could not be detected scientifically. Indeed, the pressure changes associated with storms are only equivalent to what we experience riding in a lift in an office block. And since medical literature has yet to chronicle any reports of worsening arthritis as a result of riding in a lift, the jury is still out on this one.

Another obstacle to pinning down the relationship between weather and health is the sheer number of possible atmospheric conditions. The cause of the increased pain could be attributable to barometric pressure, temperature, humidity or precipitation. Furthermore, different patients report different circumstances. Some say the pain precedes a change in the weather; others say that the two coincide; and still more say that the pain follows the change. No wonder relatively few scientists have tackled the subject.

To muddy the waters further, post-Hollander experiments have been less supportive of a link between arthritic pain and the weather. A 1985 study involved 35 people with osteoarthritis and 35 with

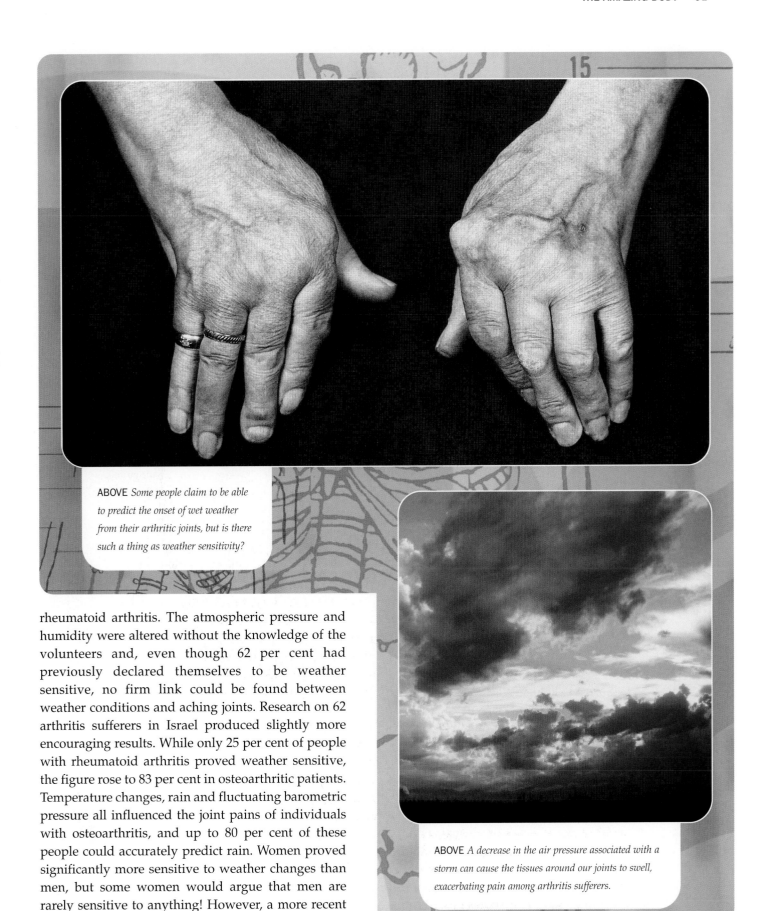

ABOVE *Some people claim to be able to predict the onset of wet weather from their arthritic joints, but is there such a thing as weather sensitivity?*

rheumatoid arthritis. The atmospheric pressure and humidity were altered without the knowledge of the volunteers and, even though 62 per cent had previously declared themselves to be weather sensitive, no firm link could be found between weather conditions and aching joints. Research on 62 arthritis sufferers in Israel produced slightly more encouraging results. While only 25 per cent of people with rheumatoid arthritis proved weather sensitive, the figure rose to 83 per cent in osteoarthritic patients. Temperature changes, rain and fluctuating barometric pressure all influenced the joint pains of individuals with osteoarthritis, and up to 80 per cent of these people could accurately predict rain. Women proved significantly more sensitive to weather changes than men, but some women would argue that men are rarely sensitive to anything! However, a more recent study by Francis Wilder, director of research at the

ABOVE *A decrease in the air pressure associated with a storm can cause the tissues around our joints to swell, exacerbating pain among arthritis sufferers.*

Arthritis Research Institute of America, failed to uncover any meaningful connections between arthritis and weather changes. But Wilder remains upbeat. "I think it's possible science hasn't caught up with the anecdotal evidence," he says.

Even assuming that there is a connection between weather and pain, it may not necessarily be physical. Instead it could be psychological. People tend to feel more miserable on wet days, and their gloomy mood may make the pain more difficult to bear. Alternatively, a rainy day may encourage elderly people to stay in bed or in a favourite armchair for longer periods, the inactivity causing their joints to feel stiffer. Sceptics also point out that if you want to believe in something badly enough, you can make

it happen. Some aches and pains can be psychosomatic. American meteorology professor, Dennis Driscoll, says: "If you convince yourself that there is a relationship between the weather and your pain, then by golly, there is one. As the barometer lowers and the clouds approach and the wind picks up, if you think that your arthritis ought to be acting up, it will."

Although there are conflicting views as to what may cause a link between wet weather and arthritic pain, there is one area on which most of the experts are agreed: don't bother moving to a drier climate – the stress involved with re-location may exacerbate the symptoms and the likelihood is that, after the first few months, your body will acclimatize to the new weather pattern and feel no better than before.

ABOVE *In tests conducted on arthritis sufferers, women's joints have proved markedly more sensitive to changes in the weather than men's.*

Why Can't we Tickle Ourselves?

Tickling is no laughing matter – at least not if you're trying to tickle yourself. Because no matter how ticklish you are when someone else runs their fingers across the soles of your feet, if you do it yourself it elicits no reaction at all. This fact has baffled neuroscientists for decades.

Scientists have long held the view that tickling evolved in early man as a defence mechanism, in order to alert the body when foreign, and potentially dangerous, objects were touching it. This helps to explain why we are ticklish in our most vulnerable spots, such as the stomach, which covers a number of major organs, and the neck, which houses the vital jugular vein.

Down the years, occasional studies have been conducted to investigate this ticklish subject and it has been determined that the reason we can't tickle ourselves is because something in the brain can

LEFT *Tickling is thought to have evolved in early man as a defence mechanism. This is why we are ticklish in our most vulnerable spots.*

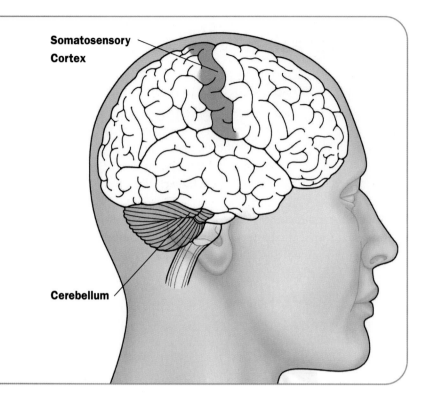

Somatosensory Cortex

Cerebellum

ABOVE *If we try to tickle ourselves, part of the cerebellum sends out a signal to dampen activity in the somatosensory cortex, the region of the brain where tickling is perceived.*

anticipate the effects of our actions and dull the sensation. Called a corollary discharge, it is sent out by the brain just before tickling is perceived and has the immediate effect of putting a dampener on any feelings. In effect the brain can distinguish expected sensations, such as when we tickle ourselves, and unexpected sensations, such as when somebody else tickles us.

Every order for the execution of a movement leaves a copy of itself circulating in the brain as a means of informing other brain areas and preparing for the movement that is about to take place. Thanks to this copy, we can perform complex movement sequences, monitor our own movements and even recognize them as our own. The copy of the motor command, however, almost certainly has a second function: forecasting and suppressing sensations that will result from the movement itself. By blocking sensations caused by the body's own movements, the brain is free to receive unexpected, and much more informative, sensations. This is why we don't feel our shoes rubbing against our soles at every step, but we do feel it when a stone gets into our shoe. It is also

why we don't feel our tongue wiggling inside our mouth with every word we say, but we do feel it if we accidentally bite our tongue. The brain acts as a censor, weeding out the obvious and saving its attention for more important things. When we read a sentence in a book or a newspaper, our eyes flicker back and forth, but the brain censors our vision during these rapid-fire movements so that we do not become queasy and disorientated. And it is this ability of the brain to sort the wheat from the chaff that explains why we can't tickle ourselves.

To investigate the matter further, three London neuroscientists – Professor Chris Frith, Dr Daniel Wolpert and Sarah-Jayne Blakemore – conducted a series of tactile tests at the Institute of Neurology in 1998. They used a robot that was ordered to tickle on command and a group of 16 volunteers who tried to tickle themselves in the name of scientific progress. After the robot had tickled the subjects' right palms with a small sponge, the subjects did the same themselves, but reported that their efforts produced a significantly less tickly sensation than that created by the robot. Then, instead of tickling one hand with the other, their movements with the left hand were transmitted to the right via the robot, after a delay so that they were effectively tickling themselves. When the delay was fleeting the sensation was minimal, but the longer the delay the more ticklish it felt. This supported the notions regarding the timing of the corollary discharge, indicating that the discharge is less effective in a longer delay between the do-it-yourself tickle and the actual sensation. By introducing a delay of just one-fifth of a second, DIY tickling was shown to be fully effective.

The next task was to attempt to locate the source of the discharge and thus the brain's censor. Some of the volunteers were placed in an MRI machine and told to lie on their backs with their eyes closed. A device with a piece of soft foam on a plastic rod was activated so that it tickled the volunteers' left palms. Sometimes the subjects operated the rod themselves; on other occasions the researchers did it from outside the scanner. The researchers compared activity in different parts of the brain during the various experimental situations so that they could ascertain which parts of the brain control tickling movement and which actually create the ticklish sensation. The results showed that during self-tickling, a region at the back of the head, called the cerebellum, is the censor that spoils all the fun. The cerebellum is the region of the brain concerned primarily with maintaining posture and balance and co-ordinating movement, but these tests indicate that its duties extend even further. During self-tickling, part of the cerebellum, known as the anterior cerebellar cortex, sends a killjoy signal to dampen activity in the somatosensory cortex, the region of the brain where tickling is perceived. However, when the researchers did the tickling, this part of the cerebellum was less active. The somatosensory cortex is the part of the brain that helps interpret external stimuli. When someone else tickles us, it signals the rest of the body to react. But when we tickle ourselves, it receives the message from the cerebellum to ignore the resulting sensation and the tickle is short-circuited.

Only one question remained: why do we laugh helplessly when we are tickled? Some have insisted that it is a reflex reaction; others, including Charles Darwin, argued that it is the result of close physical contact with another person. But the findings of the London researchers support the former theory – we laugh when we are tickled purely as a reflex.

Meanwhile, a group of Swedish scientists have made another startling discovery in relation to tickling. By comparing the brain response to actual tickles and to the anticipation of tickles, Martin Ingvar and his team found that in both cases the activity level in the somatosensory cortex was much the same. Thus they concluded that the threat of a tickle feels just like the real thing. But naturally it doesn't work if you try it on yourself…

BELOW *By comparing the brain response to actual tickles and anticipated tickles, Swedish researchers have discovered that even the threat of a tickle can provoke a reaction.*

What Makes People Sleepwalk?

It has been estimated that 10 per cent of humans sleepwalk at least once in their lives. Somnambulism, as it is also known, is most common among children, affecting around six per cent of youngsters compared with only two per cent of adults. In the majority of cases sleepwalkers calmly get out of bed, wander around aimlessly for a few minutes, and then go back to bed without causing harm to themselves or anyone else. Occasionally, however, there can be more alarming consequences.

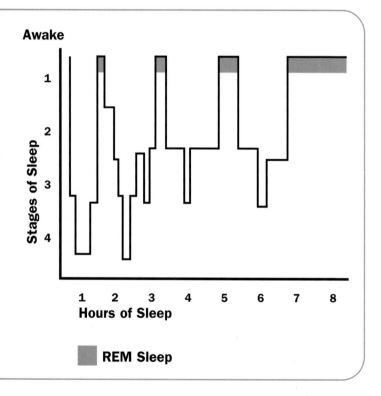

ABOVE *During sleepwalking, the area of the brain responsible for consciousness stays asleep while that which controls movement and sensory systems becomes alert.*

Sleepwalking is a genetic disorder and one that, for reasons unknown, affects men more than women. Recordings of electrical impulses produced by the brain during sleep reveal that there are two distinct types of sleep – REM (rapid eye movement) and NREM (non-rapid eye movement). Throughout the sleep period, these two types alternate in cycles. Sleep starts with NREM sleep, which makes up about four-fifths of the sleep time in adults. It begins with drowsiness, the brain waves become increasingly deep and slow until brain activity and metabolism fall to their lowest level. Dreams rarely take place during NREM sleep. About an hour and a half into sleep comes the first period of REM sleep. During this time the brain becomes more electrically active, the eyes move rapidly and vivid dreaming is most common. The first REM period may last no more than ten minutes but they increase in length as sleep continues, the final one lasting up to an hour. Sleepwalking usually occurs during deep NREM sleep early in the night or during arousal from that type of sleep. The fact that it occurs in the course of the dreamless phase dispels the familiar myth that sleepwalkers are simply acting out their dreams.

The root of sleepwalking resides in the brain. The cortex of the brain, which is responsible for consciousness, stays asleep while the area of the brain that controls our movement and sensory systems goes into awake mode. "In essence," says Dr Irshaad Ebrahim, medical director of the London Sleep Centre, "the switch mechanism from one stage of sleep to another doesn't work and we go into what is known as a 'hyper-arousal state'. We are unable to pinpoint exactly why that is, but we know it's down to complex chemical circuits in the brain."

At that point, people tend to leave the bed and become active in a confused and disorientated state. Their eyes are open, although glazed over, and they invariably move slowly and clumsily. Some talk and not only walk around, they get dressed, negotiate stairs, cook, eat, or even urinate in unsuitable places, such as cupboards. The more adventurous leave the house, get into the car and drive off down the road. In 2005 a pub landlord escaped a drink-driving conviction because he was asleep when he climbed into his BMW and wrapped it around a tree. On a lighter note, a woman recently discovered her sleepwalking husband mowing the lawn, naked!

Sleepwalking was once thought to be purely psychological, but it is now understood to be a complex combination of psychological and physiological factors as well as chemical interference. With children, somnambulism is most common among boys between the ages of four and 12, especially when they are over-tired. Night terrors – where the child wakes up screaming in a semiconscious state – and bed-wetting are also thought to be associated with sleepwalking. When an episode arises from a night terror, the sleepwalker's behaviour is more frantic and may involve thrashing around and running into walls. Fortunately most children grow out of sleepwalking once they have passed puberty. Among adults, sleepwalking has been found to occur as a side effect of excessive alcohol and certain drugs, both of which affect the chemical balance of the brain. It may also be related to stress and anxiety (something as trivial as an argument earlier in the day) and with the use of sleeping pills. In general, the sleepwalker has absolutely no recollection of the episode, which can last anything from a few minutes to nearly an hour. People who start sleepwalking as adults tend to have the problem for the rest of their lives and chronic sleepwalkers have been known to tie themselves to their beds in a bid to prevent a recurrence. Although no single neurochemical problem underlies all adult cases, Carl Hunt, director of the National Center on Sleep Disorders in Bethesda, Maryland, says a substantial number go on to develop Parkinson's disease, suggesting that sleepwalking may be part of a progressive neural condition.

Adult sleepwalking is more serious, as it is often more aggressive. Sleepwalkers are not allowed in the US armed forces because of the threat they pose to themselves and others when allowed access to weapons. An illustration of the potential threat posed by a sleepwalker occurred in 2003 when Jules Lowe, from Manchester, UK, fatally attacked his 82-year-old father, Eddie. The attack occurred while Jules was asleep and he had no recollection of the incident. Dr Ebrahim carried out a series of sleep studies on the defendant before the trial and confirmed that he had been sleepwalking. "Mr Lowe had a history of sleepwalking," he said, "and this was generally worse when he drank alcohol, but he had never been violent before. However, his stepmother had just died and

ABOVE *Sleepwalker Jules Lowe was acquitted of murdering his father on the grounds of insane automatism. He had no recollection of carrying out the attack.*

there were several other stressful factors in his life." At the time of the attack Mr Lowe was in a state called automatism, meaning he acted involuntarily. He was acquitted of murder on the basis of insane automatism, his sleepwalking having been brought on by causes within (such as stress) rather than by external factors (such as drugs or alcohol).

In another high-profile case, rock guitarist Peter Buck was acquitted of attacking BA staff on a transatlantic flight to London in 2002. The court accepted he had no recollection of the incident because he was suffering from non-insane automatism at the time, brought on by combining alcohol and a sleeping pill at the start of the flight. Ironically, Buck is a member of the band REM.

A popular misconception is that it's dangerous to wake a sleepwalker. Actually, it is very difficult to

ABOVE *REM guitarist Peter Buck (second left) was acquitted of air rage because he was suffering from a condition brought on by combining alcohol and a sleeping pill.*

wake a sleepwalker because they can't hear a word that is being said to them, even though they may be chattering away. Instead the most sensible option is to guide them safely back to their bed.

A recently discovered phenomenon, linked to sleepwalking, came to light at a 2004 meeting of the Australasian Sleep Association. Sleep physician Peter Buchanan described how a patient of his, a respectable, middle-aged woman with a steady partner, used to leave the house while she was sleepwalking and had sex with complete strangers. The woman was totally unaware of her double life until her partner, having found unexplained condoms scattered around the house, became suspicious and finally caught her in the act one night. The woman's condition was diagnosed as a case of sleep sex, a sleep disorder that is also known as REM behavioural disorder.

Normally, when we enter the REM phase of sleep in which we often dream, our bodies are immobilized, but in the case of sleep sex this doesn't happen and we can act out our dreams. Because we don't lose our muscle tone, we can actually do whatever we are dreaming about. And if what we are acting out fits with the dream, we might not wake up. British sleep expert Neil Stanley explains: "If you are lying there dreaming about having sex with your wife and you just happen to be having sex with your wife, then there's nothing there to stop the dream. You don't perceive that as wrong."

Indeed, people have committed (sometimes violent) sexual acts on themselves or their partners, while asleep. One man broke two fingers in the process of trying to escape from restraints he had made to restrict his unwanted nocturnal behaviour. US scientists studying the phenomenon found that instead of passing through the usual phases of sleep, each of which has a defined brain wave pattern, patients affected by sleep sex displayed unusual patterns during one of the phases or short interruptions in their sleep. The sleep sex behaviour took place during these hiccups in the sleep cycle. A number of the patients studied also had a history of sleepwalking. Like sleepwalking, the condition can be genetic or it can be triggered by alcohol or stress, but the US scientists noted that each of the patients also had emotional problems, without which their sleep disturbance could have manifested itself as sleepwalking or simply talking in their sleep.

5 SURGEONS AT WORK

The history of surgery dates back to around 8000BC in Ancient Egypt, where finds have included evidence of brain surgery having been carried out on a labourer. Carvings dating to 2500BC describe surgical circumcision, while operations such as castration and amputation are also believed to have been performed by the Egyptians. Unsurprisingly, many ancient practices have fallen by the wayside, but others, such as trepanation where a hole is drilled into the skull to relieve abnormal cranial pressure, have been adapted by modern surgeons. Among the first modern surgeons were those who displayed their skills on the battlefields of the Napoleonic Wars, often combining this work with their main jobs as barbers. Nowadays surgeons' jobs are more clearly defined as they find themselves at the cutting edge of modern medicine. Barely a month goes by without an innovative surgical procedure being performed somewhere in the world. In particular, the development of transplants (the first kidney transplant was carried out in the 1950s) has opened up a whole new field of surgery, enabling patients to recover the use of organs and, more recently, limbs through operations that would hitherto have been considered impossible. With each step forward, surgeons are also devising safer methods of conducting these operations. Thanks to new techniques, such as microsurgery, ultrasound and laser surgery, lives that would once have been ruined or lost can now be saved and the recipients can look forward to a long and productive future.

Boy's Arm Reattached After Shark Attack

As the sun was setting at around 8.30pm on a warm summer day in July 2001, eight-year-old Jessie Arbogast from Ocean Springs, Mississippi, was playing happily in knee-deep water some ten yards offshore at Langdon Beach, northwest Florida. He had gone to the beach for a family day out with his sister, brothers, cousins and his aunt and uncle, Diana and Vance Flosenzier. Although his sister had ventured out much further, Jessie and the other youngsters preferred to crouch in the shallow surf. Suddenly one brother felt something brush past his leg and Jessie saw the fins of a shark protruding menacingly above the water. Before he could react, the shark had sunk its razor-sharp teeth into Jessie's right arm. Vance Flosenzier was on the beach when he suddenly heard a scream and a cry of "Shark!". Turning toward the sea, he immediately noticed a spreading pool of blood in the water near where his daughters and Jessie were playing. Then he saw a 200-lb (91-kg), 7-ft (2.13-metre) long bull shark about to roll away, its huge jaws clamped on Jessie's arm. Instinctively, Flosenzier rushed into the water and grabbed the base of the shark's tail, rendering it powerless, and began to pull hard. On the second tug, Jessie was freed, but only because the shark had completely severed the boy's arm, midway between the shoulder and elbow. A large chunk of his right thigh was also ripped off. As Jessie was released from the shark's jaws, he fell back into the arms of another swimmer who saved him from drowning.

An unconscious Jessie was quickly carried ashore, but even in that short time he had lost so much blood that there was no more flowing from the gaping wounds. One witness described his thigh as resembling a chicken drumstick with a large bite taken out. His aunt used beach towels as tourniquets and tied them tightly around his arm and leg and placed T-shirts over the bone sticking out from the arm stump,

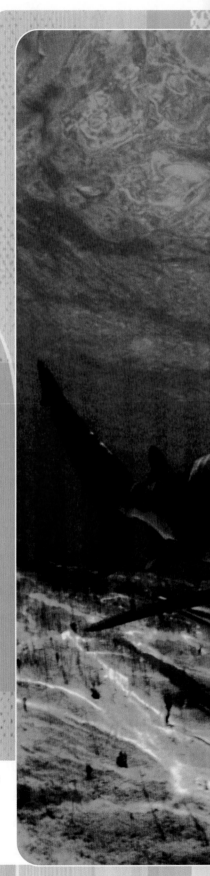

RIGHT *A 7-foot (2.13-metre) long bull shark bit off the arm of Jessie Arbogast while he was playing in shallow water off the Florida coast.*

thereby slowing the loss of what little blood remained in his body. Then, with the help of fellow holidaymakers, she began administering prolonged cardiopulmonary resuscitation. Her husband used his mobile phone to call the emergency services and soon a helicopter from Baptist Hospital in nearby Pensacola touched down. The initial prognosis for Jessie was bleak. He had no pulse and was clinically dead. His body had been drained of blood, the worst possible scenario in a trauma case. In such instances, fewer than one per cent of victims survive. One of the helicopter paramedics said: "He was white as a ghost, a paleness that comes only from blood loss. He was kind of like a rag doll." His eyes were open, but rolled back. Although the paramedics could have been forgiven for pronouncing Jessie dead on the spot, they kept the helicopter's engines going for a fast-action "scoop and run". With the help of the boy's uncle, they carried him to the helicopter and inside continued CPR while inserting a breathing tube. As they closed the door, after spending less than six minutes on the ground, they asked about the missing arm. Nobody knew where it was.

By then Vance Flosenzier, anxious to protect the other children in the sea, had wrestled the shark ashore. It was still thrashing about on the sand. When the paramedics inquired about the arm, a local ranger wondered whether it was still in the shark's mouth. So he shot the shark four times in the head, causing it to relax its jaws. The ranger then prised the shark's mouth open with a police baton and a volunteer fire fighter managed to retrieve Jessie's arm using a pair of forceps. Wrapped in moist towels and packed with ice, the arm, fresh from the shark's gullet, was rushed by ambulance to the hospital, arriving shortly after the helicopter carrying Jessie.

By the time Jessie landed at the hospital, he had been starved of blood for nearly 30 minutes. Because the principal function of red blood cells is to act as containers for haemoglobin, which distributes oxygen throughout the body, he had also been deprived of life-giving oxygen. He was taken straight down to a trauma room with medics continuing to perform CPR on his lifeless body. The immediate need was for a major infusion of blood and within 15 minutes, a nurse had pumped 3.17pts (1.5 l) of blood into him. After several anxious moments, they finally felt a pulse – just as the ambulance arrived with the arm. In total Jessie received more than 14 litres of blood.

Having stabilized his condition, the other consideration was the reattachment of the arm. "Amazingly it was a clean cut," said plastic surgeon Dr Ian Rogers. "You don't anticipate that shark bites will be clean cuts, but this was surprisingly good and surprisingly clean, specifically since it came out of the shark's gullet." Nevertheless, it was still a complex procedure, requiring the careful reattachment of one bone, three nerves, one artery, three veins and three muscle groups, if Jessie was to regain some use of his arm. While Dr Rogers marked the corresponding veins, arteries and nerves with sutures on the severed arm, orthopaedic surgeon, Dr Juliet De Campos, shortened the arm by removing about 1in (2.54cm) of bone so that the stump would hold a plate to keep the limb in position. She then clamped the bones together by means of two screws in the stump, two in the arm and two at the point of overlap. Next, Dr Rogers began reattaching the major nerve endings, each of which is about the thickness of an eyelash. When it came to reconnecting the veins, he had to take some veins from Jessie's leg to replace damaged vessels in the arm. The clamps were released and blood began to flow back into the arm, which Dr Rogers described as "absolutely white" and very cold. It was another half-hour of massaging the arm before the team felt any response and then all the small cuts in his forearm began to bleed. The hardest part, said the doctors, was stitching the skin back. "It was like putting a jigsaw puzzle together," Dr De Campos told *Time* magazine. Finally, after 12 hours of surgery, Jessie was wheeled into the recovery room.

However, while the outlook for future arm use was reasonably hopeful, the fact that Jessie had gone so long without blood between the shark attack and medical treatment meant there was a grave risk of damage to his organs, including his brain. At that stage, doctors had no idea of the extent of the damage caused by the massive loss of blood.

On the day after the emergency operation Jessie was strong enough to be moved to Sacred Heart Hospital, which housed the sole paediatric intensive care unit in the area. He remained in a critical condition and was put on dialysis after suffering kidney failure. "It's going to be a very tough situation to pull him through," confessed the head of the paediatric unit. "It's an unusual type of event for someone to go through full cardiopulmonary arrest for 30 to 45 minutes and survive."

Four days after the attack Jessie returned to surgery so that repair work could be carried out on his injured leg. This entailed removing dead skin and making a skin graft from pig skin. All the while, doctors remained concerned about possible brain damage. In the course of the healing process organs swell, but if Jessie's brain were to swell to the point where the pressure in his brain was greater than the blood pressure, no blood could reach the brain and he would die. Thankfully a computerized axial tomography (CAT) scan revealed no sign of swelling, and although his other organs were still not functioning properly, they were showing signs of improvement on a daily basis. Within a week of being savaged by the shark, Jessie had advanced from a deep coma to a light coma and was beginning to respond to pain, stimulation and commands – further indications of neurological progress. Furthermore, he was taken off the respirator and was able to breathe without help.

Over the next few weeks Jessie continued to make slow, but steady, progress. He emerged from the coma and began to focus on objects around him, although his degree of awareness was still uncertain. His parents, David and Claire, started taking him for strolls in his wheelchair around the intensive care unit. When Jessie was first admitted to hospital, none of the experts held out much hope that he would survive, let alone reach the point where he could be allowed home, but on August 12, 2001, against all the odds, he was released from the hospital and returned to Ocean Springs.

Nobody was under any illusions about the task that lay ahead but under the loving care of his family (David Arbogast gave up his job as a tile setter to become his full-time carer), Jessie continued to show enormous courage and determination in overcoming his horrific injuries. Visitors reported that he was able to say the occasional word and that he smiled and looked at whoever spoke to him, gradually displaying a greater awareness of events around him. By the summer of 2004 he was speaking more clearly, although he was still unable to form whole sentences. He could eat proper food without the need for a tube and was able to interact with his brothers and sisters, laughing and smiling. In the words of his aunt Diana he had grown "like a weed", so much so that only his father could manage to lift him. Although still unable to sit up on his own, he could roll and crawl on a special mattress.

It remains uncertain whether Jessie will ever regain the full use of his faculties following that fateful day on the beach. But the mere fact that he survived such an ordeal is a testament to his bravery and the skill of the surgeons.

Living with a Dead Man's Hand

In 1985 Matthew Scott lost his left hand in a firecracker accident, "the result," he admits, "of my own stupidity". The 23-year-old from New Jersey, had a prosthesis fitted and, despite being naturally left-handed, he taught himself to write with his right hand. Although the artificial limb was a considerable boon to his daily routine, its limitations were all too apparent and he never felt completely comfortable with it. However, with no alternative on the horizon, Scott reconciled himself to a life of severely restricted manual function. Then, while on holiday in Britain, his wife Dawn spotted a newspaper article detailing the ongoing research that was taking place into hand transplantation. On learning that the Jewish Hospital in Louisville, Kentucky, was ready to create America's first hand transplant patient – and only the second in the world – Scott, along with 300 others, applied to be the lucky recipient.

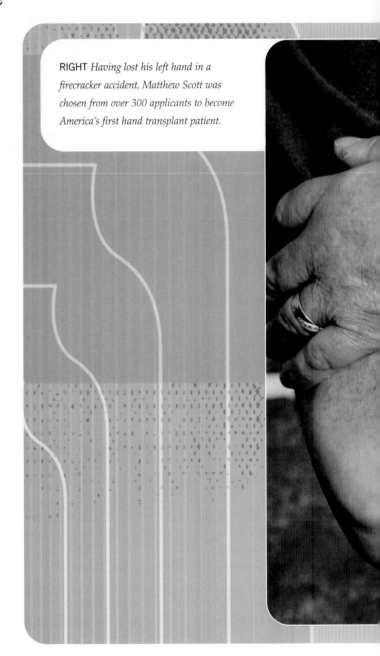

RIGHT *Having lost his left hand in a firecracker accident, Matthew Scott was chosen from over 300 applicants to become America's first hand transplant patient.*

"Don't get me wrong," he says. "Living with a prosthesis is an absolutely acceptable way to exist. I just wanted something better – flesh and bone and muscle and tendon, as opposed to plastic, latex and a battery." Before selection, he had to undergo extensive tests to ensure that he could cope with the physical demands and also the psychological challenges of wearing a dead man's hand. Eventually, on the afternoon of January 24, 1999, he underwent the transplant in a 14$\frac{1}{2}$-hour operation, with 17 surgeons working around the clock to connect the arteries, veins, nerves and bones of the dead donor's hand to Scott's left arm.

The question of hand transplants is a controversial one. The major problem with any transplant surgery is that of rejection. The body's immune system automatically launches a huge attack on the "foreign body" and while this is a vital defence mechanism in fighting infection and disease, it poses a real danger for transplant patients. Powerful drugs are used to suppress the immune system, but these can produce serious side effects, such as cancer, diabetes and high blood pressure. Although the dosage for immunosuppressive drugs administered following a hand transplant is no higher than for organ transplants (such as heart, lung, kidney, pancreas or liver), it has nevertheless been estimated that one in ten people receiving a hand transplant will die within ten years of the operation as a direct result of taking immunosuppressants. Doctors and medical ethicists have expressed concern that a patient should be subjected to the risks of surgery and an indefinite future on immunosuppressant drugs, purely for the sake of replacing a prosthetic hand with a human one. These risks are considered acceptable for organs, such as a heart or a kidney, that are essential to human life, but a hand does not fall into that category. Is it worth risking a life for a new hand? Thus critics of the

Louisville operation argued that insufficient research had been carried out into the procedure and its potential aftermath.

Despite the misgivings, Matthew Scott was subsequently hailed as the world's first successful hand transplant. Although he initially struggled to cope with the six days a week of intense physical therapy and experienced nausea and stomach upsets plus three mild rejection episodes where his body rebelled against the transplant, these eased over time. The rejection episodes were resolved by medication and the only significant complication in the first five years after the operation was arthritis in the thumb,

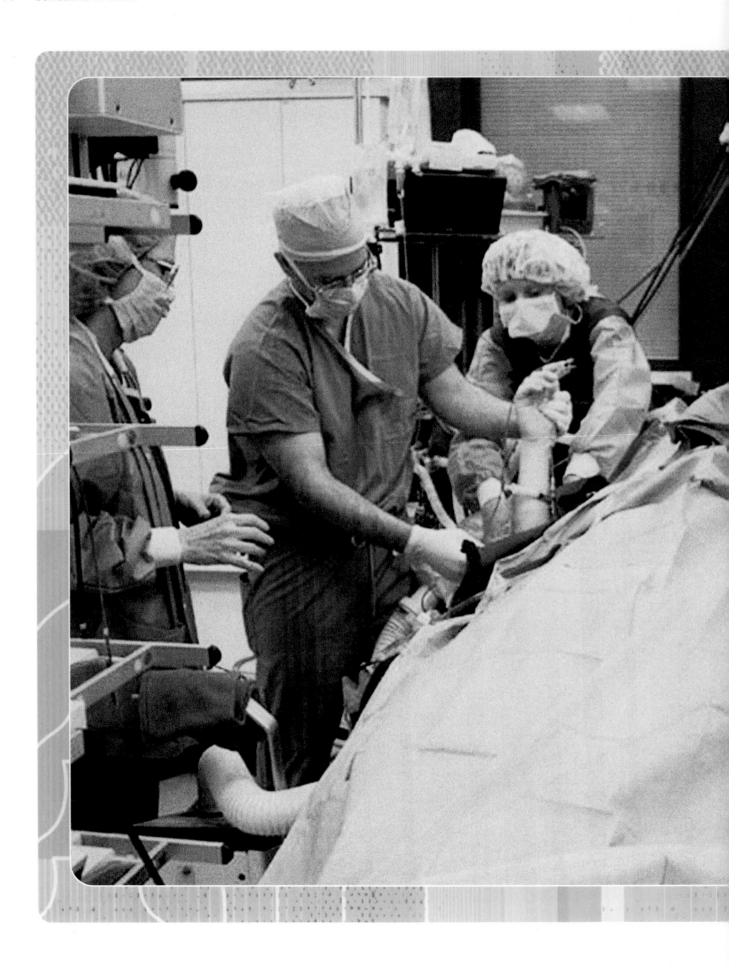

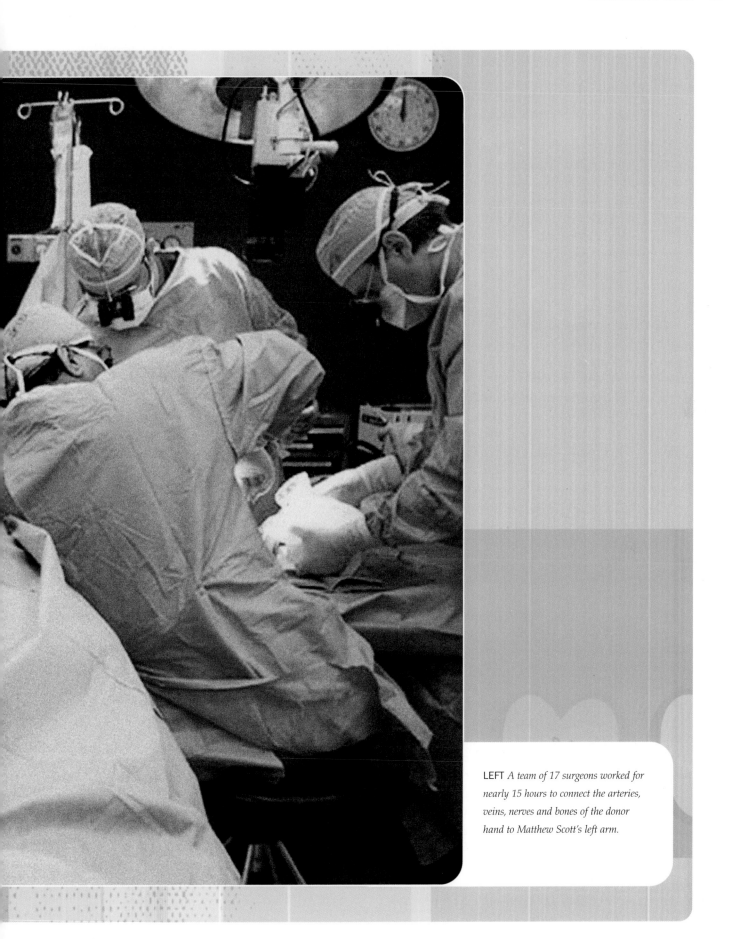

LEFT *A team of 17 surgeons worked for nearly 15 hours to connect the arteries, veins, nerves and bones of the donor hand to Matthew Scott's left arm.*

a possibility doctors were aware of before the transplant. Indeed, they say it became an issue simply because he had gained so much flexibility in the hand.

Six years on from the surgery, he was able to use the transplanted hand for such everyday tasks as throwing and catching a ball, opening doors, turning doorknobs, playing football, moving furniture, drinking from a glass, dialling a mobile phone, writing his name and tying his shoes. His annual appraisal also revealed that his pinch and strength were better and there was considerable improvement in his ability to feel different objects. He had hot and cold sensation in the left hand and could distinguish between rough and smooth or sharp and blunt textures.

"The hand transplant has taken away the anger and frustration of not being able to do things," he says. "Now I know if I can't do something, with a little more therapy, in time, I will! I've returned to doing a lot of the everyday tasks I was not able to do with the prosthesis." One of the things that has given him most satisfaction is being able to clap for his children. "There is something great and wonderful to clap for your child – also being able to hold something in my right hand and open the door with my left."

Inevitably, he still cannot do everything that he could with his original hand and will have to continue taking immunosuppressants (albeit at a reduced dosage) for the rest of his life, but he is delighted with his new limb and thinks that it has all been worthwhile, psychologically as much as physically. "Next to our voices," he says, "our hands are probably the most expressive things. So much is conveyed in a touch of a hand, and so much in the way we use our hands." His wife confirms: "He has started to do things that he loved to do – he can play the drums now. He is much happier, he is much more at peace with himself, he is much more balanced."

The head of the team that carried out the transplant, Warren C Breidenbach, said Scott had shown "a big improvement in his hand function and had a stronger thumb". He added: "This is the most successful of all hand transplants that have taken place so far, thanks to Matt and Dawn. Without the effort they put forth, this might not have been a success. His left hand is still less than a normal hand, but much more than a prosthesis."

The fact that Scott has survived without serious illness or infection is seen by proponents of transplants as proof that advances in drug therapy have markedly reduced the risk that a patient will reject a limb from an unrelated donor.

Breidenbach says:

"It is telling us that the old concept that skin was highly rejected and could not be transplanted is now incorrect. It is still true that skin promotes more response against it than muscle or kidney, but the immunosuppression is so good nowadays that we can use the same level that was used for a kidney transplant to do a hand transplant. So the drugs are good enough today to keep skin and other soft-tissue structures alive for the same risk as kidney transplant. The fact that Matthew Scott has made six years shows us that it is possible to achieve this length of time with a hand. It also implies that he has a considerable length of time in front of him that he will keep it."

However, many in the medical profession continue to remain sceptical about hand transplants. Matthew Scott's surgery was expected to open the floodgates for further hand operations but to date fewer than 30 have been carried out worldwide. Part of this reluctance may be attributable to the experiences of the world's first-ever hand transplant patient, New Zealander Clint Hallam. Hallam was given a stranger's hand in France in 1998 but found it difficult to come to terms with the new limb, the road to acceptance was not helped by the fact that some of his friends avoided him because they found it repulsive. After claiming that he felt even more handicapped than when he just had one-and-a-half hands, he asked for the new hand to be cut off in 2001 because he had become "mentally detached from it".

For Matthew Scott, however, there are no such problems. He can still remember the magical moment when he woke up after the operation to find that he had fingers on his left hand again. "Instead of a bulk of bloody bandages with nothing there, it was a bulk of bandages with fingers sticking out. I'll never forget that."

ABOVE *The world's first hand transplant – carried out on New Zealander Clint Hallam in France in 1998 – illustrated the potential perils of the procedure.*

First Human Tongue Transplant

In July 2003, doctors in Austria successfully carried out the world's first human tongue transplant during a ground-breaking 14-hour operation. The patient, a 42-year-old man whose identity was withheld, had been suffering from a malignant tumour covering the right side of his tongue, the glands, the right lower jaw and the underside of his tongue. He had been unable to open his mouth when admitted for surgery at Vienna's General Hospital and because the cancer was so advanced, surgeons had little choice but to amputate.

BELOW *Vienna's General Hospital where the world's first human tongue transplant was performed on a 42-year-old male cancer patient in 2003.*

In the past, when patients lost their tongues, a piece of their small intestine was removed and grafted on to the tongue stump. Although the small intestine is soft and able to produce mucus, therefore making it suitable for the mouth, its small size means that extra space remains in the oral cavity. Consequently, patients are never again able to speak clearly or swallow and must be fed through tubes.

By performing a transplant, the Austrian surgeons hoped to eliminate those drawbacks. The major problem facing them was in suppressing the immune system sufficiently so that the transplant was not rejected. This was a particular problem as the mouth is a non-sterile environment, due to the presence of food. However, the mouth is also very effective at keeping itself clean naturally.

Previously tongue transplants had only been carried out on animals, but the Austrian team of nine physicians, led by Dr Christian Kermer, had been waiting to try out the technique on humans. "We had been planning to do this operation for about two years," he said, "but we needed both a patient and a suitable donor. The advantage over the previous method of treatment is that the patient keeps his tongue and will be able to move and even feel it."

The tongue, from an unnamed donor, was chosen because his blood type and tongue size matched that of the intended recipient. The new tongue was first removed from the brain-dead donor by a separate team of doctors in an adjacent operating room and quickly handed over for transplantation. The donor was then taken off life support. Meanwhile Dr Kermer's team had cut out the patient's tongue through an incision from ear to ear. They then connected the muscle tissue, nerve endings, arteries and veins of the donor tongue to those in the recipient's mouth. Dr Kermer revealed that the team had attached the two nerves responsible for tongue movement from the donor organ to the patient's own nerves and had managed to attach one of the two nerves responsible for sensation.

Announcing the successful operation, Dr Rolf Ewers, another leading member of the team, said:

"The tongue now looks as if it were his own – it's as red and colourful and getting good blood circulation. The tongue is just slightly swollen. That's also a good sign, which means that probably no transplant rejection has begun. We hope that the patient will eventually be able to eat and talk normally. It's very unlikely he'll regain his sense of taste, but regaining feeling and, primarily, movement would be an optimal result. The patient was young, and removing a tongue at that age is very cruel. But the cancer was at a very late stage – he was a heavy smoker."

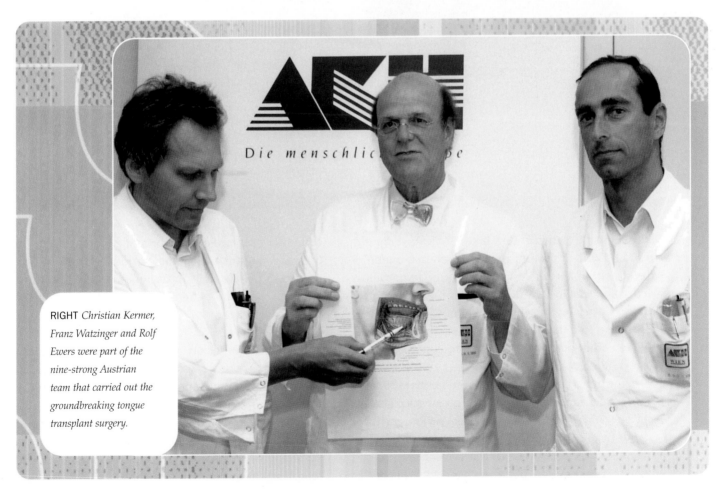

RIGHT *Christian Kermer, Franz Watzinger and Rolf Ewers were part of the nine-strong Austrian team that carried out the groundbreaking tongue transplant surgery.*

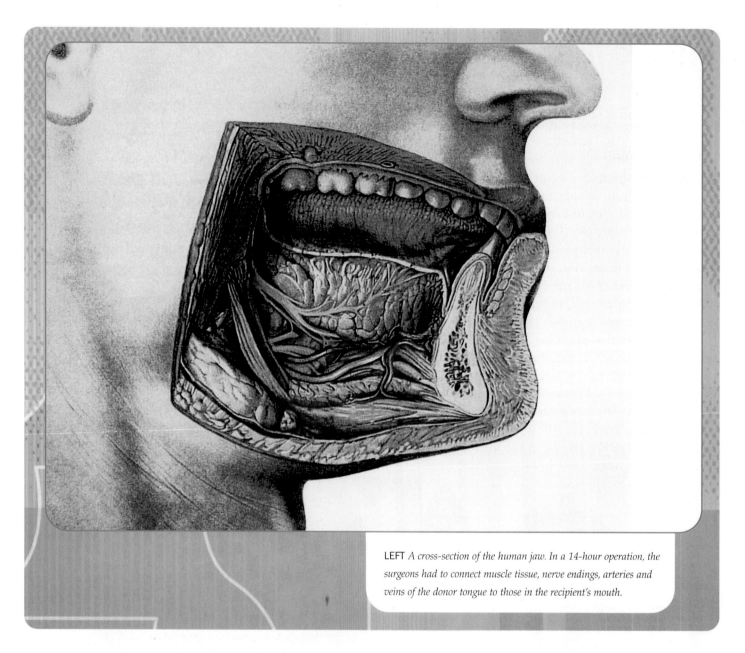

LEFT *A cross-section of the human jaw. In a 14-hour operation, the surgeons had to connect muscle tissue, nerve endings, arteries and veins of the donor tongue to those in the recipient's mouth.*

Although the patient must take medication for the rest of his life to prevent rejection, within a month of the operation he was learning to speak and swallow and could already swallow some of his own saliva and could make himself understood with the help of a tube inserted into his trachea. To help him to adjust to the surgery, he was also undergoing a course of speech therapy.

In the light of its success, the hospital plans to carry out tongue transplants whenever possible to increase the survival rate for oral cancers, which currently stands at around fifty per cent in cases of late diagnosis. It treats up to fifteen patients who could benefit from a transplant each year. Whereas in Britain people must actively agree to participate in the organ donor scheme, under Austrian law doctors have the right to use organs from any dead patient unless the patient has specifically requested not to be a donor.

However, Peter Rowe, the chairman of the British Transplantation Society's ethics committee, sounded a note of caution to those considering the surgery. "A lot of immunosuppressant therapy would be required to promote acceptance," he said. "One would have to weigh up the benefits of a transplant with the various risks of suppressing the immune system, which raises the risk of infection, and, in the long term, of further malignancies."

The Bionic Woman

A British stroke victim has become the first person in the world to have "bionic" implants fitted to help restore her lost hand and arm movement. Forty-six-year-old Fran Read from Poole in Dorset underwent the pioneering procedure, which aims to produce movement through electrical stimulation, at Southampton General Hospital in May 2005.

In 1996 and 2002 Mrs Read suffered two strokes that paralyzed her left side. She subsequently recovered almost all movement, however, apart from in her triceps and fingers. The operation was the result of a long-term experimental study by the University of Southampton in partnership with the Alfred Mann Foundation, a US medical research organization. they hoped to explore the feasibility of using radio frequency microstimulator electrical devices to improve motor recovery and re-educate arm and hand function in stroke patients. By means of a small cut, made under local anaesthetic, surgeons inserted the five, tiny, cylindrical microstimulators into her left arm, close to the nerves and muscles that had wasted away through lack of use following her strokes. Two weeks later, once the devices had settled in the patient's arm, she was fitted with a radio frequency cuff that relayed signals from a specially developed computer to the microstimulators. By pressing a button on the computer, she could send instructions to the "bionic nerve cells" in the same way that the brain usually sends messages to the muscles.

Under this procedure, both deep and superficial muscles can be individually activated allowing more selective control of movement, which is particularly useful in the forearm and hand. It is hoped that in time Mrs Read will be able to extend her elbow and wrist, open her hand and grasp objects. The strokes have not affected her ability to walk, and since she has always been a keen netball player her goal is to be able to play again. She therefore needs to be able to throw and catch a ball with both hands. If all goes well, she should eventually be able to disconnect the computer and move her arm freely as it will once again act on electrical messages from her brain.

Dr Jane Burridge, leader of the project, said that Mrs Read's movement should improve the more she uses the system. It is a question of rebuilding the muscles and re-educating the limb to recognize electrical signals linked to movement.

Dr Burridge added:

"The patient is only able to move her left arm a bit. She is able to grasp things between her finger and thumb, but has trouble releasing them. The aim is to produce a functional reaching and grasping movement. We now have to test the microstimulators to see how much we need to stimulate them to make her arm move to, for example, pick up a cup or brush her hair. Following a stroke, between 30 per cent and almost 66 per cent of patients have problems regaining upper limb function. Until now, electronic stimulation devices have not been well accepted, mainly because with surface systems people have difficulty putting the electrodes in the correct place to achieve a useful movement and implanted systems have involved major surgery. But because this system is implanted, electrodes do not need to be placed on the skin and because individual muscles are activated, a more functional, natural movement is possible. It is also less invasive than previous generations of neural implants and because the electrodes are so small they can be implanted into many different muscles, providing the potential to create the fine,

graded movement essential for hand and arm function. This is not a system to replace someone's movement – it helps to retrain their muscles to enable them to re-learn to move. If we put these microstimulators in a paralysed person it would not help them because this is really a type of therapy that can only help people who have some movement. "

Similar devices have been implanted in the arms or shoulders of patients in the USA, Canada and Japan, but these operations have only involved one stimulator. However, Mrs Read, however, had two stimulators inserted into her upper arm and three into her forearm in an attempt to co-ordinate arm and hand movement. If the system proves successful, it could also, one day, be used to help some victims of spinal cord injuries to walk again.

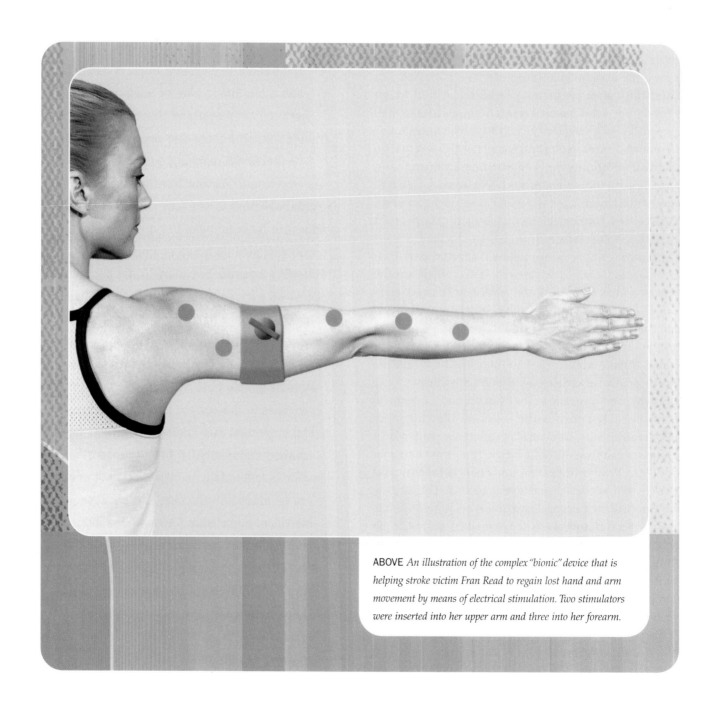

ABOVE *An illustration of the complex "bionic" device that is helping stroke victim Fran Read to regain lost hand and arm movement by means of electrical stimulation. Two stimulators were inserted into her upper arm and three into her forearm.*

Paralyzed Man Given Brain Implant

A severely paralyzed man has become the first person in the world to benefit from a brain chip that reads his mind. The revolutionary implant allows him to control everyday objects simply by thinking about them. Experts in neurotechnology hope that the implant, called BrainGate, will ultimately allow paraplegics to regain the use of their limbs.

Before July 2001, Matthew Nagle was an active sportsman. Then, in trying to defend a friend when a fight broke out after a firework display near Weymouth, Massachusetts, USA, Matthew was viciously stabbed in the neck. The knife severed his spinal cord and left him paralyzed from the neck down and confined to a wheelchair. To this day, the tip of the curved, 8-in (20.3-cm) blade remains lodged in his spine.

Unable to breathe without a respirator, 22-year-old Matthew faced a bleak future, with doctors telling him there was no chance of regaining the use of his limbs. However, technology is always looking for ways of improving the lives of disabled people. Since the 1980s Professor John Donoghue, an expert in neurotechnology at Brown University, Rhode Island, has been studying how the brain translates thoughts into action. Having worked out how neurons become "excited", his next task was to translate the electronic impulses into commands that could operate a computer or a machine. Initial experiments were carried out by implanting electrodes into the brain of a monkey that had been taught to play computer games with a joystick. Impulses from the monkey's brain allowed the animal to move a cursor on screen. Encouraged by this success, Professor Donoghue was ready to test BrainGate on humans in the hope that it might help disabled people to be more independent by tapping into their brain waves.

In June 2004, during a three-hour operation at New England Sinai Hospital in Massachusetts, Matthew Nagle became the first person to be fitted with BrainGate. A hole was drilled in his head and the aspirin-sized BrainGate chip was implanted, 0.039in (1mm) deep into his brain, just above the sensory motor cortex where the neural signals for controlling arm and hand movement are produced. One hundred wafer-thin electrodes attached to the chip are able to detect the electrical signals generated by his thoughts and then relay them through wires into the computer, which analyses the brain signals. These signals are then interpreted and translated into cursor movements, enabling him to control the computer by thought alone.

The first test came after a three-week recovery period following the operation. Matthew was shown a cursor moving on screen and told to think about the direction in which it moved. The computer attached to the chip in his brain recorded the impulses he had when thinking about the cursor moving left, right, up and down. Each direction was associated with a characteristic pattern of signals from his brain. Then the computer was programmed to recognize each pattern and to move the cursor accordingly, for example if he thought "down" the cursor moved down. Although Matthew is unable to move his limbs, he has learned to move the cursor around the computer screen simply by imagining moving his arm. The computer screen is basically a TV remote control panel, and in order to indicate a selection he merely has to pass the cursor over an icon. When he goes over that icon, it is the equivalent of a click on a mouse. Consequently he has been able to perform tasks such as opening his e-mail, which previously would have been beyond him. "We're essentially providing a way of connecting his brain to the outside world," said Tim Surgenor from Cyberkinetics, the company that makes BrainGate.

By using software linked to devices around the room, Matthew has since been able to think his TV on and off, change channel and alter the volume. He has also developed the ability to use thought to open and close an artificial prosthetic hand and move a robotic arm to take sweets from one person's hand and drop them into another. He can even draw computer art and play computer games such as Pong and Tetris.

For some time medicine has used neural implants that respond to commands. Neurosurgeons insert electrodes into the brain that use electric signals to alleviate chronic pain and relieve the symptoms of Parkinson's disease, epilepsy and depression.

Professor Donoghue hopes that BrainGate will greatly improve the quality of life for people with severe disabilities, perhaps enabling patients to move wheelchairs, use the Internet, and control lights, the telephone and other devices, all with their thoughts. The ultimate goal is that one day they might regain some use of their limbs. "If we can find a way to hook this up to his own muscles, he could open and close his own hands and move his own arms. We're very encouraged by Matthew, but we're cautious. It's just one person. There's further to go, but we're absolutely on the way."

Matthew Nagle is looking still further ahead, to the day when he might even be able to walk again. "It has changed my life," he says. "And I don't care if I have to use a cane, I'm going to walk. It will happen in a few years, I know it."

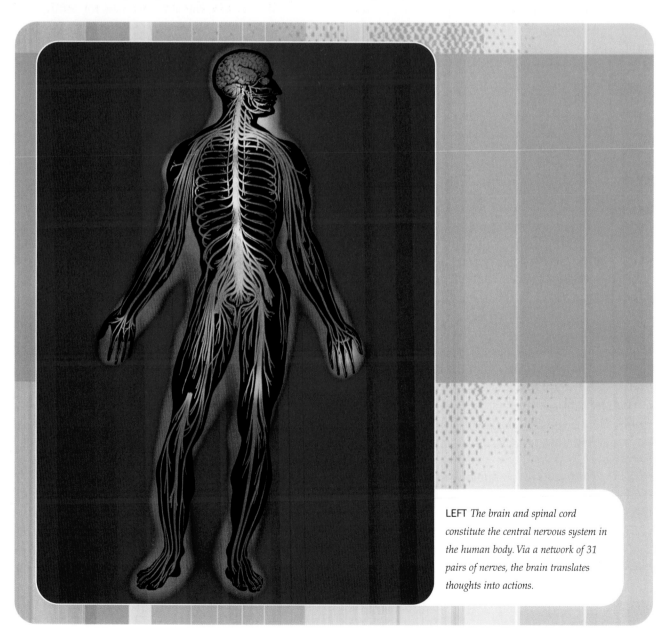

LEFT *The brain and spinal cord constitute the central nervous system in the human body. Via a network of 31 pairs of nerves, the brain translates thoughts into actions.*

Stem Cell Transplants

Of all the issues in modern medical science, few are more contentious than stem cell transplants. Indeed, many people immediately associate the very words "stem cell science" with an immoral, unacceptable act because of the misconception that all stem cells are harvested from embryos. Pro-life groups reject the practice of cultivating embryos for the express purpose of harvesting stem cells, and this has encouraged scientists to examine the possibilities of creating nerve stem cells from adult tissue instead.

Before we are born, embryonic stem cells manufacture the 200 types of other cells that every human is made of. After birth, adult stem cells repair damaged cells in the body. The aim of regenerative medicine, which endeavours to heal injuries and cure diseases by harnessing the body's own ability to heal itself, is to harness the work that these cells do naturally in order to treat other conditions – often that they were not necessarily designed to heal.

Stem cell science is still in its infancy. While embryonic stem cells, by virtue of their ability to develop into any of the 200 possible cell types, do appear to be considerably more versatile than adult stem cells, but there are worrying drawbacks associated with their use, quite apart from the ethical consideration. For example, the use of embryonic stem cells in mice has occasionally led to the development of large tumours. Such risks are greatly reduced with the use of adult stem cells, which have the additional advantage of being able to be harvested from a patient's own body, as and when they are needed. The goal now for scientists is to discover which areas of the body can be used most effectively in these procedures.

In fact adult stem cells have been used in operations for some time. Bone marrow transplants are carried out regularly in hospitals and in essence these are stem cell transplants, in that the cells used satisfy the definition of stem cells – in other words, they last a lifetime and are able to regenerate nerve tissue. An example of the true potential of adult stem cell transplants is the case of Kim Gould, an English woman who, in May 1998, was left paralyzed by a riding accident. Her horse stumbled at the final jump of a cross-country course and catapulted her into the air.

"The way I landed just snapped my spine," she said. "When they come up to you and say you're never going to walk again, you just think: 'Oh, once I get out of hospital I'll be fine.' It doesn't really sink in. In the first couple of years after my accident I just felt I wanted to sit indoors and mope. To be stuck indoors and restricted was awful. My whole way of life just changed completely."

Mrs Gould explored all avenues of treatment before taking part in a pioneering trial in Lisbon in which stem cells were transplanted from her nose into her spine. The nine-hour operation was carried out at the Egaz Moniz Hospital in October 2003 by Dr Carlos Lima, who was building on the work conducted in the late 1970s by Pasquale and Ariella Graziadei at Florida State University.

They found that there is one part of the nervous system, a region in the nasal cavity concerned with the sense of smell, where nerve fibres are in a state of continuous growth throughout adult life. This is important because we lose them all if we have a cold, but at the same time we don't lose our sense of smell permanently – that returns. From this we can deduce that those nerve cells are capable of being replaced from a stem cell. This is one part of the nervous system where there is an ongoing replacement and formation of new nerve cells throughout our lives. Because nasal tissue contains stem cells, which last a lifetime and are able to regenerate nerve tissue, it also offers a potential method of patching up a broken spinal cord.

BELOW *Paralyzed by a riding accident, Kim Gould took part in a pioneering trial in which stem cells were transplanted from her nose into her spine.*

RIGHT *Following the operation, Kim Gould began to feel more sensation. Within a year she was able to crawl around the floor and now stands on her own.*

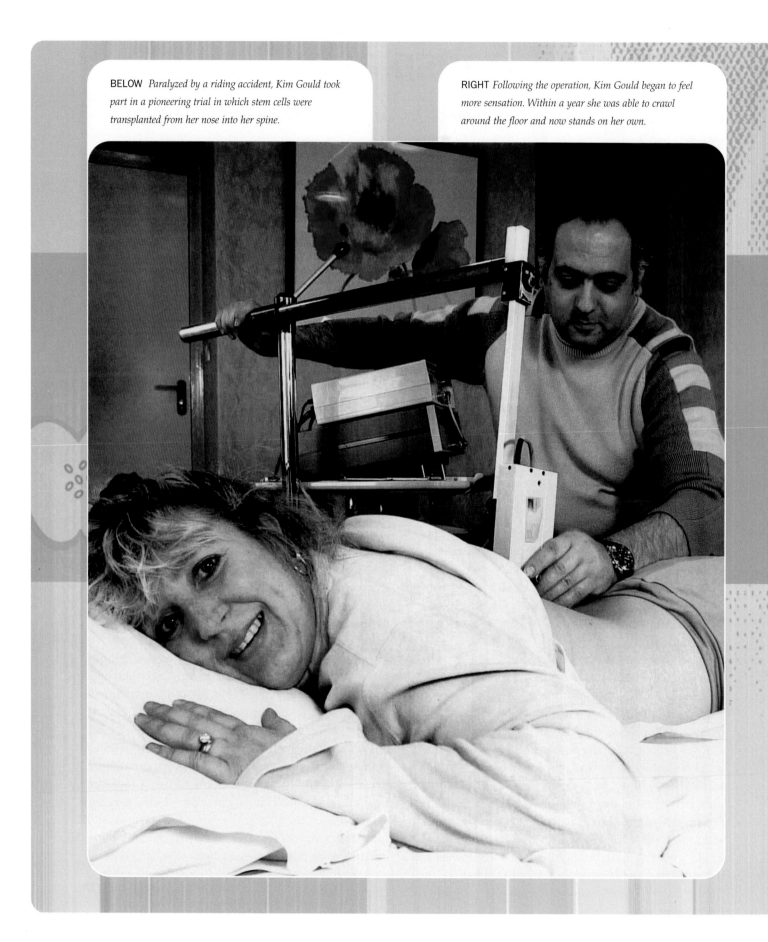

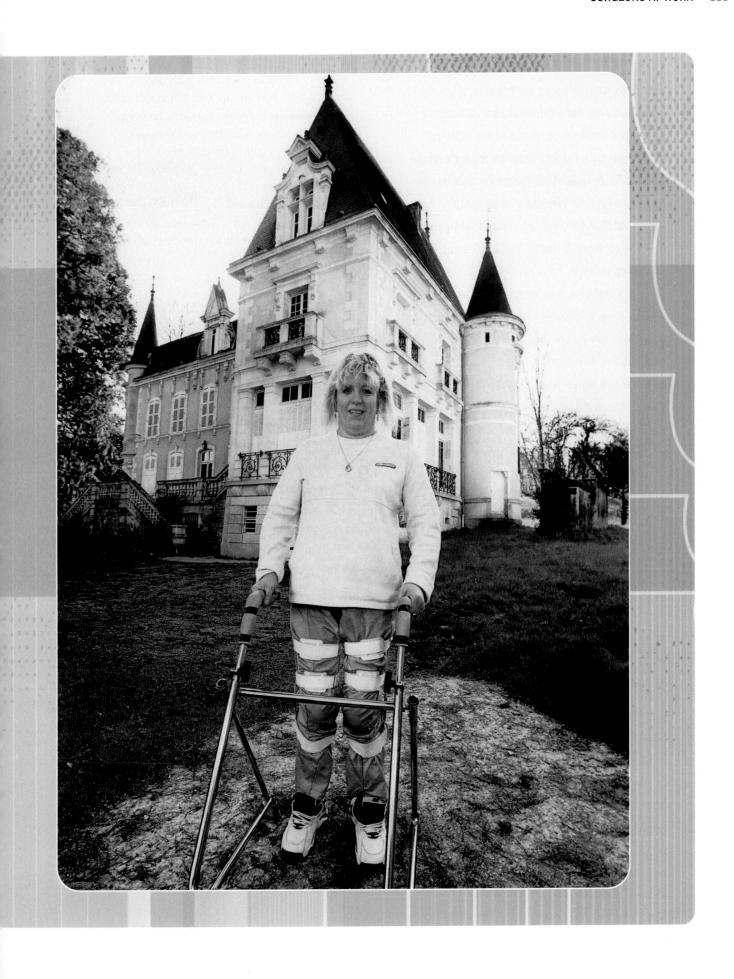

Dr Lima said:

"I am opposed to using embryonic stem cells, but not only for ethical reasons. Mother Nature made embryonic stem cells to proliferate and adult cells to replace and repair. To defy Mother Nature's laws is, at least, dangerous. Whereas here, Nature does most of the job, not us. I am asking the patient to treat himself because when you put the cells in your spinal cord, they're yours, it's natural, you've got an environment that is a very friendly one, and so the cells will grow. We know that the stem cells will stay active for months, even years. So we hope to get more function even years after the surgery. All of our patients have some kind of recovery. We have no doubts about sensory recovery and some voluntary motor recovery. They move and feel below the lesions as never before. And there is even some bladder and bowel control recovery."

Following the operation, Kim Gould began to feel more sensation down her right side, then a difference in her lower back and abdominal muscles. Within a year, she was able to crawl around the floor.

"I am quite balanced now,'" she said, "and I can actually lift each leg and move it forward. What I have recovered in a year, after six years of no movement, is quite remarkable. There's always a risk if you have cells from somebody else, or from a foetus, that they'd be rejected. But with regenerative therapy it's encouraging your own cells to regenerate and to repair yourself. If there's a chance that you could possibly be better than being stuck in a wheelchair, paralysed, I think you've got to take it."

What makes her recovery all the more impressive is the fact the olfactory tissue used for the transplant can diminish over time, making patient age an important consideration; at 43, Kim Gould was the oldest patient to participate in the trial.

Another of Dr Lima's patients is Joy Veron, a Texas schoolteacher who was injured by a horrific accident in October 1999 ,while trying to save her young family during a roadside stop near a canyon in the Colorado Rockies.

She remembered:

"My children jumped in the front seat before my mother and I made it to the car. And they were all three standing in the front seat. As the car lurched and lunged forward, I remember them all looking at me with these huge eyes, and we immediately all took off running. It was headed for the canyon. I could see the kids going over in the car in my mind as I was running. And so I got in front of the car and tried to stop it with my hands, which, of course, I couldn't. And I remember feeling it begin to run over me. I fell backward and the front of the car caught me by the heels. I did, like, a somersault under the car and the third time it hit I knew that it broke my back. At that point I went flat, and the rear tyre rolled over the length of my body. They tell me the only thing that saved my life was that I had my head turned, and the tyre missed it. I was 30 years old and the whole life I knew was gone."

Although her father managed to grab the handbrake, thus saving the lives of her

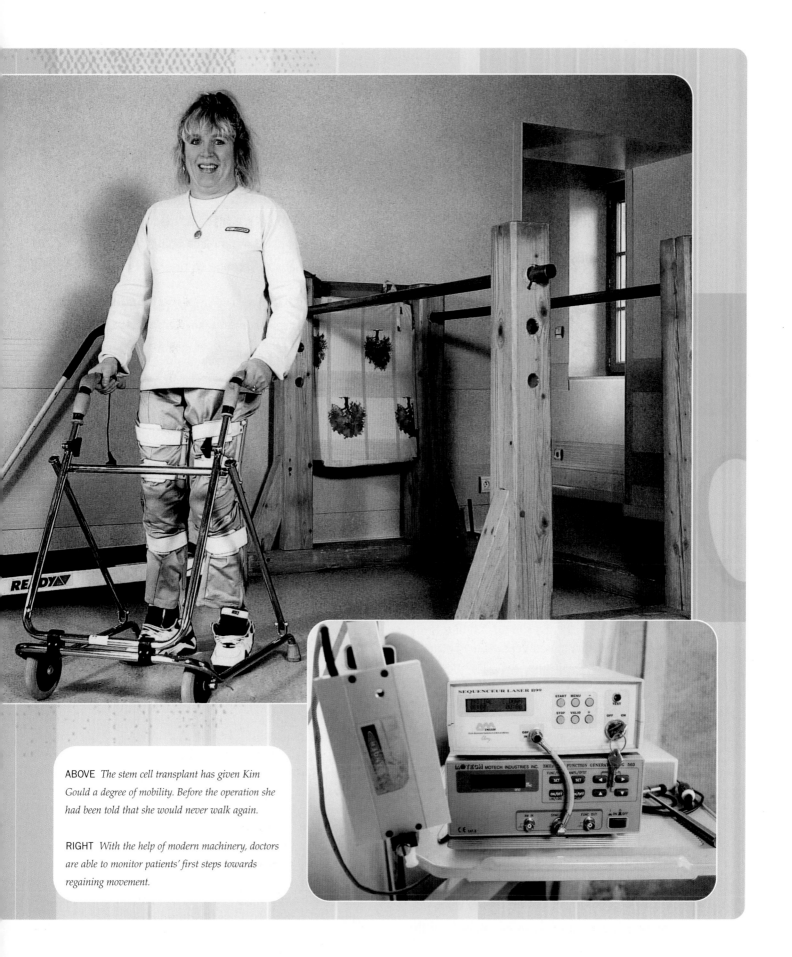

ABOVE *The stem cell transplant has given Kim Gould a degree of mobility. Before the operation she had been told that she would never walk again.*

RIGHT *With the help of modern machinery, doctors are able to monitor patients' first steps towards regaining movement.*

children, Joy Veron was unfortunately left paralyzed from the waist down.

Nine months after the operation, she was able to report a degree of improvement, although perhaps progress has not been as great as she had hoped. "The biggest improvement has been in my left leg. It was icy cold; the right was always warmer and stronger. Now the left is stronger than the right, and that is hugely significant. I'm also feeling more pain, but pain receptors are the first to regenerate, so it's good." She knows there is still a long way to go. "Of course, my goal from the surgery would be to walk again and be as I was before, but I told Dr Lima that anything I didn't have before, any little bit, was gravy for me."

One man who has been taking a close interest in Dr Lima's results is Professor Geoffrey Raisman, director of the Spinal Repair Unit at University College London. For he, too, has been researching the possibility that stem cells from one's own nose could offer a safe and effective therapy for spinal injuries. In one trial on laboratory rats, Professor Raisman's team cut the nerves that led to the animal's left forepaw. With the injury, the animal could no longer use the paw to climb effectively, nor could it use that paw to reach for its food. Stem cells were then harvested from the rat's nose and transplanted around the damaged nerve. Within weeks, the treatment produced noticeable results. "We've been able to restore the ability to climb," reported Professor Raisman. "We've been able to restore complex reaching and control movements of the use of the forepaw – the sort of functions that a patient would want to recover if they didn't have the use of their hand." When Professor Raisman then transplanted cells from the nose into the injured spinal cord of rats, the results were equally impressive. They found that the cells had a remarkable capacity to integrate into the damaged pathways and lay a "bridge" over the gap in the nerve fibres caused by injury.

"When we transplant cells into that area of damage," added Professor Raisman, "the function comes back. You're seeing a glimpse through a doorway that has never been opened before. This is what will get people out of wheelchairs. This is what will make stroke patients get better. This is what will restore the optic nerve in blindness and the auditory nerve in deafness. If we can push that door open, there's an immense amount behind it. This will be revolutionary, if successful."

He stressed that he doubted if full recovery would ever be possible.

"But if a person can't even move their arms, to be able to throw a switch, to be able to manipulate a piece of machinery, to be able to drive a car – it's going to make an enormous difference to their way of life."

Meanwhile, in South Korea scientists were acclaiming a South Korean woman who, having been paralyzed for 20 years, was able to walk again after they repaired her damaged spine using stem cells derived from umbilical cord blood. Hwang Mi-Soon had been bedridden since damaging her lower back and hips in an accident two decades earlier but was able to walk into a press conference in November 2004 with the help of a walking frame. "This is already a miracle for me," she told reporters. "I never dreamed of getting to my feet again."

For the transplant – believed to be the first of its kind in the world – stem cells were isolated from umbilical cord blood, which had been frozen immediately after the birth of a baby and cultured for a period of time. The cells were then directly injected into the damaged area of the spinal cord. Within two weeks the patient could move her hips and after a month her feet responded to stimulation and she began to take small steps with the aid of a walking frame. Doctors were amazed by the speed of her progress, but agreed that more research is needed. Han Hoon, president of government-backed umbilical cord blood bank, Histostem, acknowledged: "Technical difficulties exist in isolating stem cells from frozen umbilical cord blood and finding cells with genes matching those of the recipient." Unlike embryonic stem cells, there is no ethical problem to consider, and it has also been found that umbilical cord stem cells trigger little immune response in the recipient. So although more research is still required, and many more tests need to be carried out, could nasal and umbilical cord blood stem cell transplants be the way ahead for thousands of patients for whom there had previously seemed little hope?

The Man Whose Head was Reattached

The most amazing thing about Marcos Parra's story is that he is alive to tell it. In 2002 the 18-year-old was driving in Arizona when his car was hit by a drunken driver – the impact was so severe that Marcos's head was technically severed from his body. Only skin and his spinal cord kept it connected. He was rushed to the emergency room at St Joseph's Hospital in Phoenix where doctors admitted that they had never seen such terrible injuries. The fact that he had a broken clavicle, pelvis, tailbone and ribs was almost immaterial: what stunned the medics was the state of his neck. Normally the skull is attached to the neck by a thick ligament that runs from the base of the skull to the first vertebra. During the accident his head was thrown forward with such force that it ripped his ligament and opened up a space between the head and the neck. There was a wide separation of the bone where the skull was torn from the cervical spine, leaving his head detached from his neck.

Such injuries would usually be fatal or, at the very least, permanently crippling, but Marcos had the good fortune to be treated by Dr Curtis Dickman of the Barrow Neurological Institute at St Joseph's. By consequence it happened that Dr Dickman had been perfecting a technique to treat injuries as rare as Marcos's and had actually been testing the method on human cadavers. Marcos ended up being the first person in the world to undergo the experimental surgery that saved his life.

Because the spinal cord and arteries had not been damaged, Marcos was an excellent candidate for Dr Dickman's technique, which involved the use of two surgical screws. Surgeons delicately inserted the screws through the back of his neck to reconnect the first vertebrae to the base of the skull. This pulled the bones back into position. Then a piece of Marcos's pelvis was used to patch his neck and skull together.

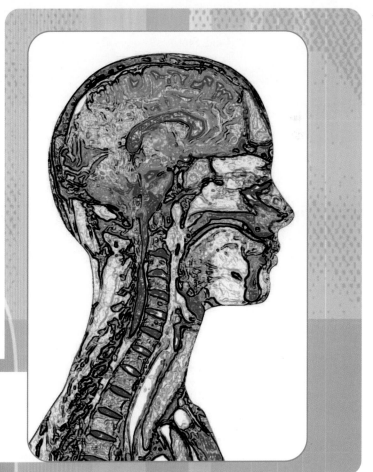

RIGHT *Whereas the skull is usually attached to the neck by a thick ligament, Marcos Parra's injuries were so bad that the ligament was ripped, leaving his head detached from his neck.*

LEFT *Dr Curtis Dickman who perfected the technique of inserting screws through the back of Marcos Parra's neck to reconnect the first vertebrae to the base of the skull.*

Afterward Dr Dickman said: "Most people with an injury like that die at the scene of the accident because it requires very severe and violent forces to create that type of injury. What distinguishes this technique from the other techniques is that it preserves the majority of motion in the neck. For when the motion is lost, it's disabling."

Following the surgery, Marcos spent four months wearing a halo-brace around his head to hold his head and neck still and to help his neck heal. He also went through hundreds of hours of rehabilitation.

The operation was such a success that, as Dr Dickman had promised, he only lost about five per cent of his neck's range of motion – an astonishingly small amount considering the seriousness of his injuries. In fact, within a few months he was even back on the basketball court and enjoying every minute of it. Unable to remember much about the accident, Marcos was just happy to be alive and to be able to be walk rather than face the prospect of spending the rest of his life in a wheelchair. He was quite simply a very lucky young man.

New Jaw Grown on Patient's Back

A German man, whose lower jaw was removed because of cancer, was able to eat his first solid meal in nine years courtesy of a replacement jawbone that had been incubated in a muscle on his back. The operation, performed at the University of Kiel in 2004, was the first in which doctors used a patient's own body to incubate and synthesize a substitute for lost bone tissue. The new jaw was grown on a titanium frame just beneath the patient's shoulder and was created using a combination of computer aided design and bone stem cells. The pioneering technique had previously been performed on pigs.

The 56-year-old patient underwent surgery to remove cancerous tumours on his jawbone in 1995 and between then and the transplant could eat only soft foods and soup. However, within four weeks of receiving his new functional jaw, he was able to tuck into a meal of sausages and bread.

In similar cases doctors can sometimes replace a lost jawbone by cutting out a piece of bone from the lower leg or from the hip and then chiselling it to fit into the mouth. Flat areas, such as the shoulder blade, are often used, but these are unsatisfactory when attempting to replicate a complex three-dimensional structure such as the jawbone. Also, bone grafts are invariably painful and slow to heal, causing loss of bone density in the donor area plus a secondary site of possible infection. In this instance the patient was taking the blood thinning drug Warfarin for an aortic aneurysm, as a result of which doctors considered it too dangerous to take bone from another part of his body in case of post-operative bleeding. Instead they bypassed the initial bone removal procedure by growing the required bone from stem cells in the patient's own bone marrow.

The Kiel medical team, led by Dr Patrick Warnke, began by making a three-dimensional scan of the patient's mouth, from which they created a virtual jawbone on a computer. The information was used to create a scale model made of Teflon on to which was fitted a thin titanium mesh cage. The Teflon was then removed to leave a perfectly sculpted, hollow, U-shaped tube. Next, this cylinder was filled with bone

mineral blocks of bovine origin, blood from the patient's own bone marrow containing stem cells, and

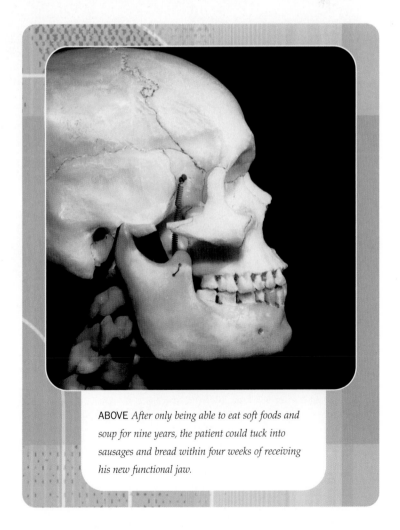

ABOVE *After only being able to eat soft foods and soup for nine years, the patient could tuck into sausages and bread within four weeks of receiving his new functional jaw.*

LEFT *In the Kiel operation, the replacement jawbone was attached to the stumps of the patient's original lower jaw and the structure's blood supply was connected to his existing jaw muscle.*

a human growth factor called bone morphogenic protein (BMP). The mineral blocks served as a scaffold, while the BMP coaxed stem cells inside the blood to form a new bone. The replacement jaw needed to develop its own blood supply so it was grafted into a part of the body rich in blood vessels – the latissimus dorsal muscle beneath the patient's right shoulder. "He didn't find this at all uncomfortable," Dr Warnke told *New Scientist*, "and was able to sleep on that side with no problems." The patient was only given standard antibiotics to prevent infection.

Doctors monitored the replacement mandible's development, and CT scans indicated that new bone was forming. After seven weeks of growth, the graft was removed, along with a flap of muscle containing blood vessels. In a three-hour operation it was then attached to the stumps of the patient's original lower jaw and microsurgery was used to connect the structure's newly developed blood supply to his existing jaw muscle and to blood vessels at the base of his neck. The skin was closed around the new jaw as much as possible. "It was a perfect fit," said Dr Warnke a few weeks after the operation. "He is very happy with the result. He can chew food again and his speech is much more easy to understand, especially on the phone." Within nine weeks of the operation the patient was able to eat steak although it had to be cut up for him because he had no teeth to bite through the meat. Assuming that bone density continues to improve, Dr Warnke hopes to remove the titanium cage and shape and smooth the new bone before implanting false teeth.

Implanting the cage into the patient's muscle meant his own tissue developed around it. "Because it was his own tissue, we don't expect any problems of rejection," added Dr Warnke.

Three Limbs Reconnected After Horror Accident

What started out like any other children's birthday party ended in a world first as surgeons attempted to reattach both arms and one leg of a ten-year-old boy after they had been severed in a freak accident.

Terry Vo was playing basketball at a friend's house in Perth, Western Australia, in March 2005 when he decided to demonstrate a slam-dunk. But as he did so, the brick wall of the garage, to which the hoop was attached, collapsed. The combination of sharp brick edges and steel guttering sliced through both arms, about 2.36in (6cm) above his wrists, and through his left leg, between the knee and the ankle. Although he had lost a lot of blood and must have been in extreme pain, Terry remained remarkably calm, showing no emotion. As the ambulance was called, his friends collected the severed body parts, which were then put in an ice bucket. When the ambulance arrived, he was still conscious and alert, and asked the woman driver: "Am I the worst accident victim you have ever seen?" She replied that since most people only lose one limb at a time, he definitely was.

At Perth's Princess Margaret Hospital for Children, surgeon

ABOVE *The wall of a garage collapsed on ten-year-old Terry Vo during a game of basketball at a friend's house in Perth, Australia, in March 2005.*

Robert Love quickly assembled three teams, involving eight surgeons and 18 other medical staff including nurses. Each team worked separately on one of the limbs. First, the body parts were cleaned of all dead tissue and brick dust. Then a steel plate was prepared for each of the two main bones in both of his forearms – the radius and the ulna – these were held in position with screws to stabilize the forearm for surgery. A similar treatment was carried out on the severed leg. Because Terry's limbs had been so badly crushed, the bones were shortened by between 1.18in (3cm) and 1.58in (4cm) before being re-attached to the body. Meanwhile, the major nerves, tendons and blood vessels were identified and deep tendons under the radial artery were repaired to re-establish blood flow. This was done by microsurgery using sutures that were finer than a human hair. The remaining tendons on the front and back of the hands were repaired, although each hand had lost at least 20 tendons. Despite the patient being a triple amputee, the operation was completed in just six and a half hours, time being of the essence lest the severed body parts died. The following day, skin from Terry's right thigh was used to graft on to open wounds.

When Terry woke from sedation, after what is believed to be the first time anywhere in the world that surgeons had re-attached three limbs to one patient in a single operation, the medical staff were heartened by the fact that he could move his thumbs. Although he was unable to feel hot and cold or pinpricks at that stage, because of the damage that had been done to his nerves, everyone was optimistic about his long-term prospects.

Sadly, despite the operation initially being pronounced a success, a few days later Terry's left leg had to be amputated, 5.51in (14cm) below the knee, when it emerged that the reattachment had not worked. The muscles in his foot had died.

"Blood was still flowing to the foot," explained Dr Love, "but the muscles had died in such a way that the foot died from the inside out. We had the ability to try and increase the blood supply to the foot but what we were beaten by in Terry's case was the fact that the inside of the foot was dead. That would lead to the small muscles in the foot actually constricting, the toes bending over, and the end result being a foot that is clawed over and doesn't have good sensation.

Even if you can get all of that to survive, he would be worse off than having an amputation. What is very disappointing is that for the first two days after the operation the foot looked absolutely magnificent."

Terry took this latest setback with typical stoicism. Dr Love told reporters: "He saw the need for the foot to be amputated and was far more accepting of it, and far more in favour of it, than the staff or the parents. When I asked him how he felt about the prospect of having an amputation, he said, 'a little bit happy, and a little bit sad'."

Nevertheless, the overall prognosis for Terry remained positive. The muscles in both hands were still alive, the skin grafts had taken and he was soon able to move each finger 0.39–0.79in (1–2cm). Doctors said they hoped to fit a prosthetic lower limb to his left leg and, because his knee-joint is intact, he should have a near-normal walking gait. The fact that he is at an age when his body is still growing, coupled with his own determination, point the way to a bright future. Dr Love summed up: "There are a lot of people in our society that function extremely well with amputations. So while we are extremely sad about it, this young fellow is extremely positive."

ABOVE *For someone so young, Terry Vo has remained remarkably positive throughout his horrific ordeal, even when his left leg had to be amputated below the knee.*

A Very Public Autopsy

It was the hottest ticket in town. On a cold November day in 2002, over a thousand people queued in the rain outside London's Atlantis Gallery in the hope of gaining admission to Britain's first public autopsy for 170 years. The controversial event – conducted by German professor of anatomy Gunther von Hagens – attracted curious onlookers from all walks of society. When the doors closed on a sell-out audience of 500, each of whom had paid £12 for the privilege, many more were left disappointed outside.

The object of the excitement was a 72-year-old man who had donated his body to von Hagens' *Body Worlds* exhibition of human corpses. Von Hagens' fascination with the human body dates back to when he was only six. A haemophiliac, he had gashed his head and ended up spending six months in hospital. But it was seeing his first autopsy at the age of 17 that inspired him to follow a career in medicine. In the 1980s he began combining medicine with art, thanks to a technique that he had invented called plastination. In this process the natural body fluids are replaced with a solid plastic that not only preserves the tissues, but also gives rigidity, thus enabling the corpse and organs to be displayed in whatever position is required. Von Hagens himself intends to be plastinated when he dies. His *Body Worlds* exhibition of preserved human corpses and body parts in various stages of dissection opened in

RIGHT *The controversial German professor Gunther von Hagens who, in November 2002, carried out Britain's first public autopsy for 170 years.*

Japan in 1995 and has proved so successful that it has made its creator a multi-millionaire. In the first seven months of opening in London – from March 2002 – it attracted over half a million visitors. The public autopsy was connected to the exhibition, the idea being that the corpse's organs would be taken back to Germany after the post-mortem to be plastinated and form part of *Body Worlds*.

However, von Hagens faced a chorus of opposition to his plans. The British Association of Clinical Anatomists warned that to take a post-mortem out of licensed premises and perform it in a public place had ethical and moral implications. HM Inspector of Anatomy, Dr Jeremy Metters, sent him a letter informing him that the autopsy would be a criminal offence under the Anatomy Act as neither von Hagens nor the venue had post-mortem examination licences. Scotland Yard also told the professor that the show could be illegal, but von Hagens remained defiant, insisting that it was educational and that he was making a stand for the "democratization of anatomy". He added that he would rather go to jail than cancel the event. In the end Scotland Yard compromized by asking anatomy professors to attend and monitor proceedings.

The autopsy was shown on giant screens inside the gallery. Flanked by a Channel 4 television camera crew, who were recording the event for later transmission, von Hagens, wearing his trademark fedora, and his colleagues informed the audience that the corpse was that of a businessman who, after losing his job at the age of 50, had embarked on a daily intake of 60 cigarettes and two bottles of whisky. He had died in March 2002. After a lengthy preamble, in which von Hagens explained that the principal aim of the autopsy was to establish cause of death, he whipped off the white sheet covering the body and made his first incision – a Y-cut from shoulder to shoulder and then down to the pelvis. While a few members of the audience shifted uneasily in their seats, von Hagens proceeded to tease away first the skin and then the subcutaneous tissue before removing the thoracic shield – the bone sitting above the subcutaneous tissue – and putting it on a steel tray so that it could be taken around the auditorium by one of his assistants.

The professor diligently explained each step and fielded questions from the audience. After opening the chest, von Hagens reached in and pulled out the heart, lungs and liver. Then, with the help of a colleague, he removed the abdominal block, including the intestines, kidneys, spleen and pancreas. In total, eight organs are removed in a standard autopsy – the heart, lungs, spleen, liver, kidneys and the brain. By now the upper body was little more than an empty cavity and as the stench from the open corpse began to fill the room, members of the audience started to cover their mouths and noses. Then they gasped as von Hagens cut from ear to ear across the skull before loosening the skin of the face and placing his hand inside the cavity. Next he produced a hacksaw and began sawing into the skull in order to remove the brain. As he sawed, he revealed that performing a post-mortem required considerable strength. "The bone is quite strong," he said, "and it takes some time to go through the skull. I listen, and from the tone I know when I end." Several people covered their faces as the dead man's silvery hair was parted by the hacksaw but von Hagens, still wearing his hat, took it all in his stride.

After an interval, during which time the organs were arranged on trays in front of the body for closer inspection, von Hagens began the examination to determine cause of death. One by one, the organs were dissected physically and verbally. The lungs were shown to be inflamed, possibly as a result of chronic bronchitis, the heart and liver were about a third bigger than they should have been, there were a number of gallstones present, and there was a tumour in one of the kidneys. The calcium deposits in the aorta were shown around the audience as an example of arteriosclerosis, the deceased's history of heavy smoking having clearly hardened his arteries. With the spleen and prostate showing no signs of disease and the kidney tumour dismissed as not unexpected in a man of that age, von Hagens concluded that, from the swelling of the heart and the lungs, the man had obviously suffered from the failure of these two organs.

Although the cause of death was hardly a great surprise, most of those who attended – many of them medical students – agreed that it had been a fascinating experience. The only thing that shocked them was the lightning speed with which Von Hagens worked, but as one young lady aptly acknowledged: "There is no delicate way of doing this with finesse."

Transplant Cure for Diabetes Sufferer

In the UK alone, 250,000 people have type 1 diabetes, also known as insulin-dependent diabetes. The condition, which usually appears before the age of 35, brings misery to sufferers, but now, thanks to a groundbreaking cell plant technique pioneered in Canada, a permanent cure for this most serious form of diabetes looks to have been found.

Diabetis is when the blood sugar level in the body is too high because the body cannot use the sugar properly. This is because the hormone insulin, which enables the body to control blood sugar levels, is either not produced by the islet cells in the pancreas or does not work properly. Insulin is responsible for the absorption of glucose into cells, for their energy needs, and into the liver and fat cells for storage. If there is a deficiency of insulin, the level of glucose in the blood becomes abnormally high, causing the patient to pass large quantities of urine and develop an excessive thirst. The body's inability to store or use glucose also causes weight loss, hunger and fatigue. Type 1 diabetes develops rapidly, often between the

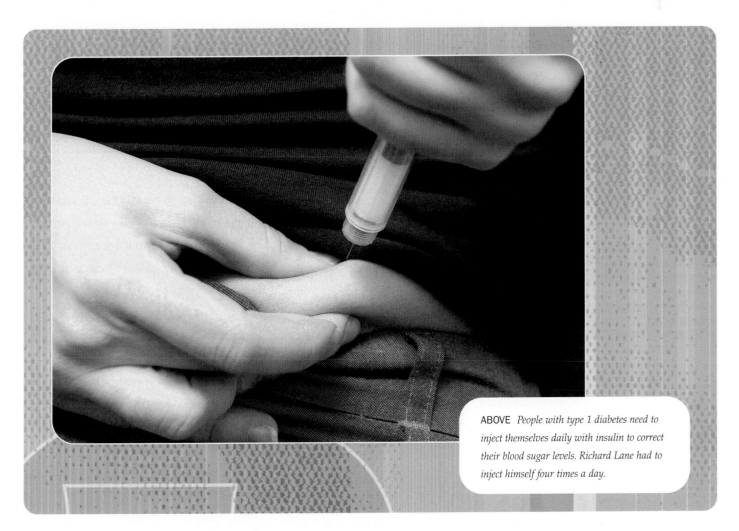

ABOVE *People with type 1 diabetes need to inject themselves daily with insulin to correct their blood sugar levels. Richard Lane had to inject himself four times a day.*

ages of 10 and 16. The insulin-secreting cells in the pancreas are destroyed by the body's immune system, most commonly following a viral infection, and insulin production ceases almost completely. Without regular injections of insulin, the sufferer lapses into a coma and dies. Type 2 diabetes has a more gradual onset and usually develops in people over the age of 40. Insulin is still produced, but not in sufficient quantity to meet the body's needs, especially when the person is overweight. People with type 1 diabetes need to inject themselves daily with insulin in order to correct their blood sugar; those with type 2 diabetes can sometimes control their blood sugar by avoiding foods that cause a rise in sugar levels or by taking medication.

The procedure that has given new hope to type 1 sufferers was perfected by British surgeon James Shapiro at the University of Alberta in Canada. Scientists had been attempting to transplant islet cells for a quarter of a century, but it was not until Mr Shapiro sharply increased the number of transplanted cells and used different types of anti-rejection drugs that positive results were obtained. Performed under local anaesthetic, the transplant involves taking healthy islet cells from the pancreases of dead donors and injecting them through the portal vein into the patient's liver. Once there, they develop their own blood supply and begin to produce insulin, the liver effectively acting as a back-up pancreas. In 2000, Mr Shapiro was able to report that seven patients had then been free of insulin injections for 11 months.

The success of the Canadian trials encouraged scientists at King's College Hospital, London, to refine the technique for growing, harvesting and transplanting the cells. The treatment, which, because of a donor shortage, was only offered to those who had major problems with conventional insulin therapy or who suffered dangerous hypoglycaemic (low blood sugar) attacks, was given to two patients in the UK – both still required small doses of insulin afterward. Then, in 2005, came the triumph that the King's College team had been hoping for.

Richard Lane, a businessman from Bromley, Kent, was diagnosed with type 1 diabetes in 1976, since then he had been insulin dependent, having to inject himself four times a day. He used to get as many as six hypoglycaemic attacks a week, some of which rendered him unconscious. In 1997 he had a serious car accident after suffering a blackout when his blood

sugar levels dropped too low, and required major surgery on his spine. In an attempt to control the "hypos", he began using a 24-hour insulin pump, which gave him a dose of insulin every six minutes. Initially, the pump had the desired effect, but then the hypos returned and complications with his eyes meant that he needed laser treatment. Retinopathy – damage to the retina, the light-sensitive area at the back of the eye – is an all too common complication for diabetics. As a result of his ill health, Mr Lane had to retire from his job as a partner in an accountancy firm. He was first offered the chance of being a human guinea pig in an islet cell transplant a few years ago, but he had turned down the opportunity. However, he later changed his mind and in September 2004 received his first transplant. The following month he had a second transplant and finally in January 2005 he had his third transplant.

In March 2005 it was announced that 61-year-old Mr Lane had not suffered one "hypo" attack since his first transplant six months earlier. After initially taking a small amount of insulin at night – three units compared to the previous dosage of 80 – to protect the islet cells, he no longer needs insulin injections, making him the first British patient to reach that all-important goal. Although he will have to take anti-rejection drugs for the rest of his life, he says it's a small price to pay. He is now able to walk briskly for 30 minutes each day and has lost a considerable amount of weight.

"I haven't felt better in myself for 30 years," he said. "It means I don't have to worry about going for a long walk. Before, if I went out for a walk without a Mars bar or CocaCola I would finish in a heap on the floor. My wife used to dread me going out of the front door in case there was a call from the ambulance service. Now I don't have to worry because it is all working naturally inside me, my body is acting as if I never had diabetes at all. It is almost like being a totally different person."

Professor Stephanie Amiel, who leads the diabetes team at King's College Hospital, is delighted by

Mr Lane's progress since the islet cell transplants. She predicts: "This breakthrough is hugely exciting. The implications for the future are enormous. Eventually this could mean the end of insulin dependence for all type 1 diabetes sufferers."

Despite Mr Lane's experience, the technique is not yet perfect. Many patients still require top-up insulin because the transplanted cells do not produce enough to control blood sugar. Furthermore, about a million of the cells are needed, which means that more than one donor is required for every transplant. Only about 800 pancreases are donated annually in Britain, which does not go far when there are a quarter of a million people waiting for treatment. The shortage of donors is clearly a major obstacle to be overcome, but it is hoped that stem cells could help make more of the islet cells or that an even newer technique,involving living donors, might provide the solution.

The latter scenario received a boost recently when Japanese

ABOVE *Following his cell transplant, type 1 diabetic Richard Lane no longer needs insulin injections. He is the first British patient to achieve that goal.*

doctors acclaimed the first transplant from a living donor. A 27-year-old woman, a type 1 diabetes patient since the age of 15, was given islet cells from the pancreas of her mother. Prior to the transplant, the daughter had been suffering hypoglycaemic attacks in which she lost consciousness every two days. Due to cultural sensitivities in Japan concerning the use of pancreatic islet cells from dead donors, the mother volunteered instead. At first it was feared that the donor herself might become diabetic because of the loss of so many islet cells, but the team at Kyoto University reported that cells from half the mother's pancreas were sufficient to free the recipient of her insulin dependency within 22 days. Two months later she was still insulin-free and her mother had suffered

no complications. The researchers claimed that the outcome was as good as that achieved with the cells of two or more whole pancreases from dead donors. They believe this may be due to the superior potency of islet cells from a living donor. While they caution that the daughter may need insulin injections in the future and that the transplant may not last more than five years, they are confident that she will be free of the "hypos" that endangered her life. As many as a quarter of people with diabetes suffer from recurrent hypoglycaemia and around 15 per cent of those cannot be improved using conventional therapy. So while the use of a live donor remains contentious, diabetics the world over will be keeping a close eye on the Japanese research.

Mandy's New Mandible

Mandy Kemp was born with Goldenhar syndrome, a rare defect that affects about one in every 20,000 births. The non-genetic, prenatal condition prevented her mandible jaw joints and other facial features from growing normally in the womb. Her jaw and chin were thrown out of alignment, creating a visible lopsidedness to her face and impairing her ability to chew, laugh, smile and talk. Her left eye was also damaged and one of her ears didn't form correctly, although she subsequently had the latter surgically reconstructed. In all, she went through 18 operations and medical procedures over the first 14 years of her life, several of them aimed at repairing the massive bone loss caused by Goldenhar syndrome. None, however, brought effective, lasting change until she met Dr Jay Selznick, a Las Vegas oral and maxillofacial surgeon, who proposed a groundbreaking implant that, if successful, would be the last major surgery that Mandy should ever need.

When Mandy was younger, other children at school would cruelly taunt her about the way she looked. "When they'd see my eye, they'd call me Cyclops," she said, "and tell me to join the circus as a freak." Mandy coped with the barbs by displaying the same fortitude that she used to deal with the syndrome itself – she never felt sorry for herself. Between her schooling came regular trips to hospital as she underwent several bone grafts in an attempt to replace the missing pieces of her jawbone. Then, in 2001, Dr Selznick informed her of his bold plan – he wanted her to become the first person in the world to receive a full prosthetic jaw.

In the standard bone grafts that she had undergone previously, only around 15 per cent of the missing jaw is usually replaced. Selznick's goal, however, was to replace the full 70 per cent of Mandy's jaw that was missing. He consulted with Dr Robert Christensen of

LEFT *Mandy Kemp endured 18 operations and surgical proceedures in the first 14 years of her life before receiving an implant to replace the missing parts of her jaw.*

a Colorado-based implants company, and together the physicians decided to replace most of Mandy's lower jaw with a one-piece, horseshoe-shaped onlay graft. The idea sounded great in theory, but when Mandy learned that it had never been tried out on a human before, she admitted to some misgivings. "I thought, 'Am I going to be a lab rat? How do they know what to do?'"

Selznick soothed Mandy's fears, and about five months before the surgery she sat for a detailed CAT scan from which Christensen created a precise, three-dimensional model of her skull and jaw. That model gave Mandy a first indication at what her new face would look like. "It looked cool," she admitted coyly. The full prosthetic mandibular implant, made of cobalt-chromium material for strength, would be a perfect fit to Mandy's skull.

In March 2002, the 15-year-old went into hospital to have her new mandible fitted. Following the six-hour operation on the afternoon of March 9, she spent the remainder of the day and night bandaged and breathing with the help of a ventilator. The following day the bandages were removed, and Mandy saw herself in the mirror. "I looked at myself and said, 'Wow. Is this what normal looks like?' I was never going to see my old face again. I was excited, and I was real emotional." Indeed, she was so emotional that she burst into tears of joy and wrote on a makeshift card: "Dr Selznick, I can't thank you enough. I will love you for the rest of my life."

While Mandy started to grow accustomed to using her new jaw, she also had to get used to her new look. "When I saw myself before I had my surgery, that was normal to me even though it was different to everybody else. Now I'm normal to everyone else, but I'm different to myself. It's very confusing. I'm going to get a different reaction from people." One positive aspect to her old disfigurement was that it "made me see that you've got to look at people and how they are on the inside. I don't judge anybody by the way they look outside – by the colour of their skin or weight or hair colour or anything. It would be hypocritical of me."

As for Dr Selznick, he said the case represented one of the highspots of his career. "It's exciting to be part of a case that is raising the standard for the medical profession," he stated. "What's even more exciting is seeing a girl who can look at herself in the mirror and feel proud of what she sees."

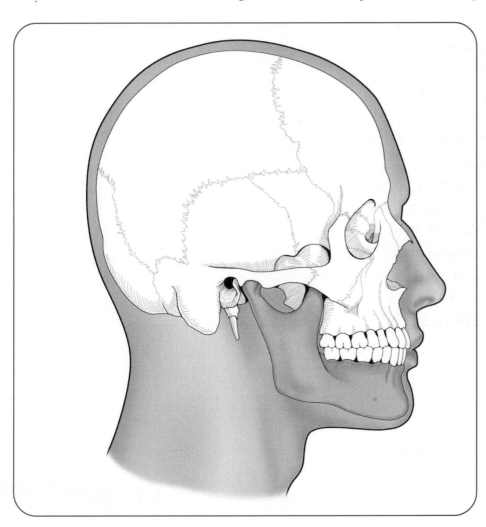

LEFT *A normal human jaw. Mandy Kemp was born with Goldenhar syndrome, a rare defect that left her face lopsided and impaired her ability to chew, laugh and talk.*

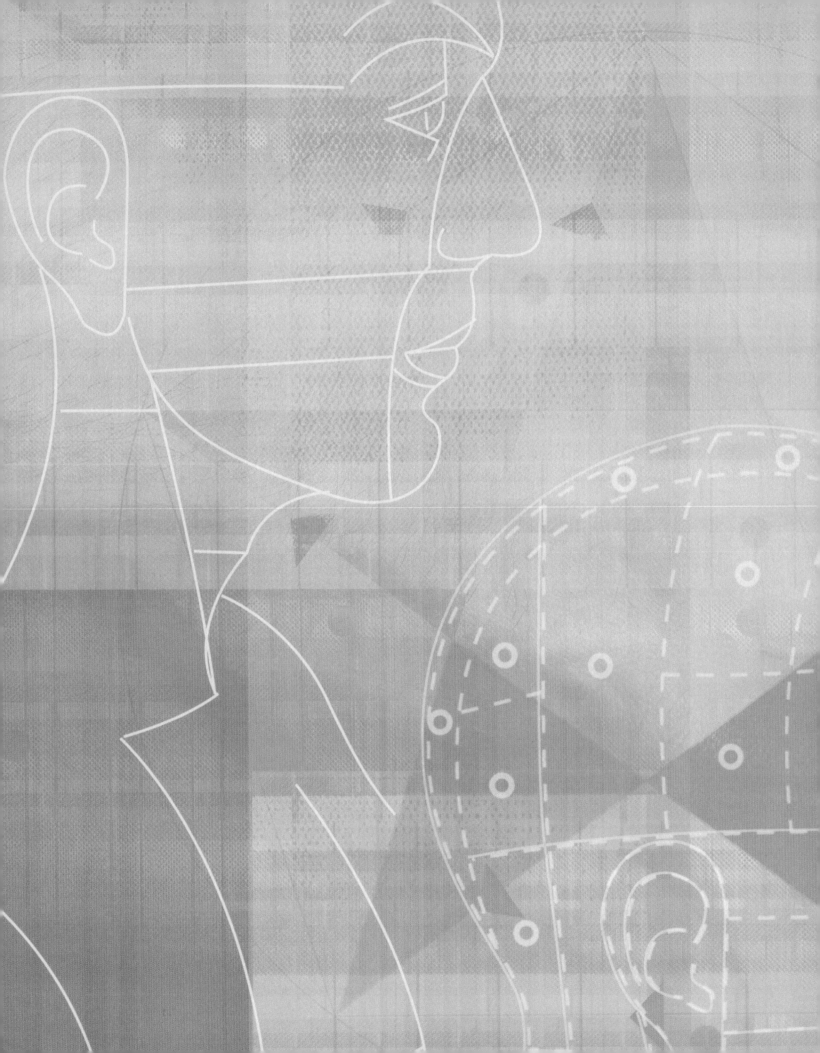

6 ALL IN THE MIND?

Of all the organs in the human body, it goes without saying that none is more complex than the brain. It receives, sorts, and interprets sensations from the network of nerves that extend from the central nervous system to every other area of the body. It has been calculated that there are more signal connections in the human brain than there would be in a telephone exchange that connected everyone in the world. We know that some brain areas are associated with specific functions. For example, muscle movements are controlled from the precentral gyrus, while touch and pressure sensation is perceived within the neighbouring postcentral gyrus. Yet many of the brain's activities remain a mystery and the secrets that it holds could go a long way to revealing the truth behind some of the most unusual and disturbing syndromes. So precisely what causes one hand to have a mind of its own, as in Anarchic Hand Syndrome? Are areas of the brain somehow linked to profound religious beliefs and out-of-body experiences? Is it possible for an organ transplant recipient to inherit personality traits from the donor? And what happens when we lose our sense of proprioception, the essential system responsible for body co-ordination?

Sympathetic Pregnancy

For centuries diverse cultures around the world have practised a ritual whereby the man took to his bed at the same time as his pregnant partner and proceeded to mimic birth by feigning contractions and birth pains in unison with the mother. In Papua New Guinea, as soon as pregnancy was discovered, the man would hide himself away from the rest of the village and build a hut, filling it with food and clothing. As the birth approached, he would stay in bed, imitating the pain of childbirth, until eventually his wife entered his new hut and handed him the baby. Similarly, the Basque males of northern Spain would, on learning of an impending birth, take to their beds and complain loudly about labour pains and spasms, thus ensuring that they received the same degree of attention from nurses as their partner. Various explanations were put forward for this behaviour – that the man's screaming drew bad spirits away from mother and child; that it strengthened the bond between father and child; that the father's simulated birth asserted his paternity; and that it was a form of anxiety relief for the man.

These rituals have always been viewed as nothing more than fraudulent simulations. A variation on the theme emerged around the turn of the seventeenth century when essayist Francis Bacon wrote of a new phenomenon – that of sympathetic morning sickness being experienced by the husband. This was such a bizarre development that it, too, was largely dismissed as apocryphal. Bacon himself offered no real explanation save for the suggestion that some husbands are so devoted and caring that something strange happens to their bodies when their wives become pregnant. Despite the lack of medical rationale, the condition did not go away. In 1878 the *Lancet* reported an instance of a husband and wife suffering bouts of simultaneous morning sickness and ten years later, the *New York Medical Gazette* told of a man who began having morning sickness two weeks after his wife missed a period – before she was even diagnosed as pregnant. Moreover, he had experienced morning sickness during each of his wife's previous pregnancies.

Sympathetic pregnancy has now been accepted as a genuine medical condition and given its own name – Couvade Syndrome, from the French word for "hatching". It applies to any symptoms that a man suffers during his partner's pregnancy. The symptoms may include weight gain, nausea, insomnia, indigestion, heartburn, fatigue, backache, persistent toothache, variations in appetite, headaches, diarrhoea or constipation, itchy skin, mood swings and food cravings. Up to 80 per cent of expectant fathers experience Couvade Syndrome in some form or another although only a handful display the more dramatic symptoms, such as stomach spasms during the actual birth. Research indicates that the symptoms are most profound during the third and fourth months of the pregnancy and then again just as the birth is imminent. All of the symptoms, however, seem to disappear immediately after the birth.

There is a suggestion that Couvade Syndrome is hereditary or that it is more likely to occur among men who have been adopted or have experienced infertility. Meanwhile, recent tests in Canada indicate that the men may be experiencing hormonal changes associated with parenthood and that these are similar to maternal changes. Numerous other theories have been put forward by psychoanalysts, ranging from jealousy of the mother's ability to carry the child to a display of guilt for having impregnated her. Some claim that the psychosomatic condition is a way of identifying with the foetus or that it is simply a

ABOVE: *Essayist Francis Bacon suggested that some husbands experience sympathetic morning sickness out of concern for their pregnant wives.*

demonstration of empathy with the partner and the changes that she is going through. The truth is that at present sympathetic pregnancy remains a baffling phenomenon but one that is becoming ever more common in the western world, with the social changes that now allow the man to take a more active and caring role during pregnancy.

Phantom Limb

Up to 80 per cent of all amputees experience a sensation whereby they can still feel their missing limb long after the surgical wounds have healed. The feeling may occur immediately after the amputation or many months, even years, later. The sensation, first described in 1866 by American neurologist S Weir Mitchell while observing Civil War soldiers, is known as "phantom limb".

The phantom limb is usually described as having a tingling feeling and a definite shape, which resembles that of the living arm, hand or leg before amputation. Amputees often feel pressure in a missing arm or leg when it is actually the stump that is being touched. The phantom limb is said to move through the air in much the same manner as a normal limb would behave when the person walks, sits down or stretches out. Initially, the phantom limb feels perfectly normal in size and shape – to the extent that the amputee may reach out for objects with the phantom hand or try to step on to the ground with the phantom leg. However, those who have experienced the phenomenon say that, in tim,e part of the limb starts to change shape, becoming less distinct and sometimes fading away completely, leaving the hand or foot apparently dangling in mid air. On other occasions the phantom limb is slowly "telescoped" into the stump until only the hand or foot appears to remain at the stump tip.

Many phantom limb sensations occur after an injury to the site of the amputation. Thus it is possible that someone who was born without a limb and who did not originally experience any form of phantom sensation could suddenly feel one as a result of some kind of injury to the stump. One such case concerned an 18-year-old girl who was born missing her left arm below the elbow. She had felt no phantom sensations until one day, while out riding, she fell from her horse and landed on the tip of her stump. Thereafter she developed a feeling that she

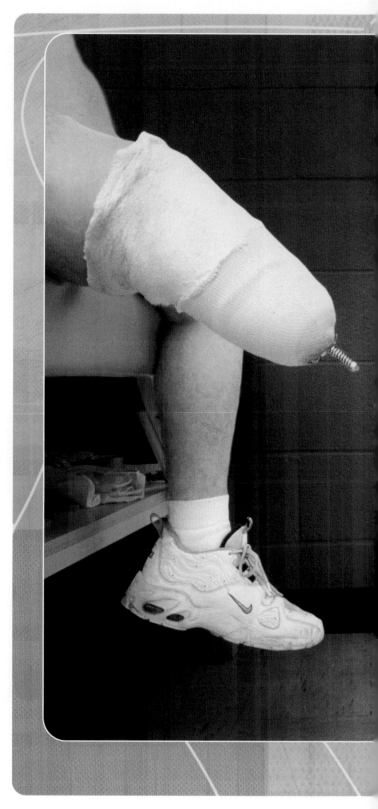

had a full-length phantom arm, hand and fingers. The sensation, which was described as pleasant and painless, persisted for about a year before disappearing altogether.

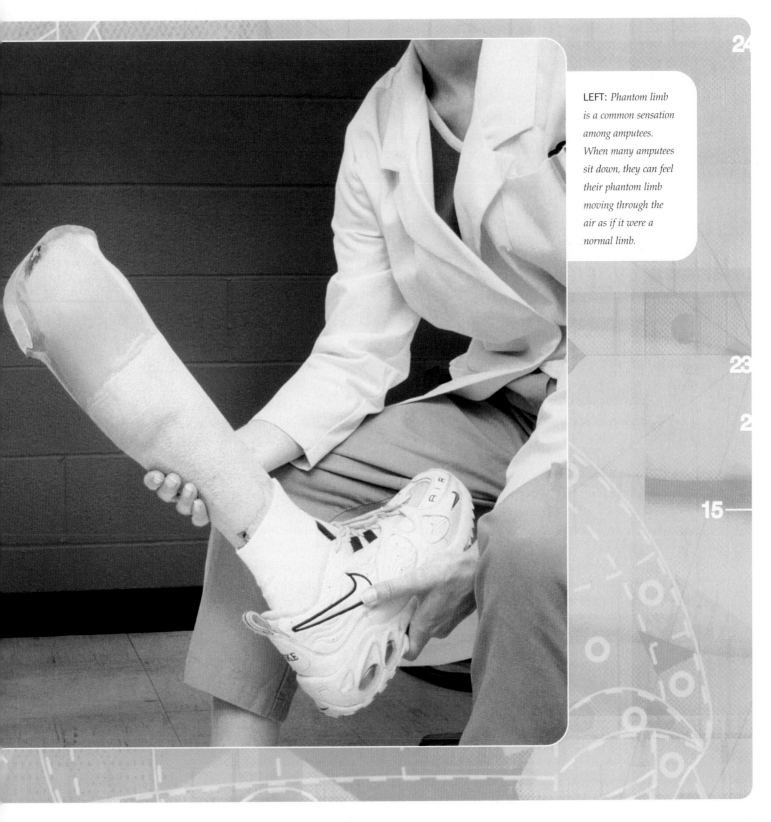

LEFT: *Phantom limb is a common sensation among amputees. When many amputees sit down, they can feel their phantom limb moving through the air as if it were a normal limb.*

In another case study, a 15-year-old girl, who lost a leg to cancer, kept a log detailing any examples of phantom limb sensation. On the first day after surgery she felt an itching and tingling sensation in the area where her toes had been. On the second day she said that it felt as if her entire foot was, asleep but the sensations were relieved when she massaged her other, non-amputated, foot. Each sensation lasted around ten minutes. After ten days the sensations began to disappear and by the end of the month they had vanished entirely. However, some phantom feelings persist for years.

So what causes phantom limb? Research suggests that perception of our limbs is "hard-wired" into our brain and that sensations from the limbs become mapped on to these brain networks as we develop. In many instances the image of the functioning limb has been stored in the brain since childhood and this image persists after the limb itself or the use of it has been lost. Usually the problem disappears over a period of time, as the patient corrects his or her body image. Although, as we have seen, some people who are born with a missing limb or who lose a limb under the age of four can later experience phantom limb sensation. It usually only occurs in the event of an injury to the stump, simply because the image of the whole body never had the time to be imprinted on the brain.

In recent tests to investigate the phenomenon, scientists from University College London measured brain activity while volunteers took part in an experiment with a rubber hand. While the volunteers hid their right hand under a table, a rubber hand was placed in front of them to make it appear like part of their body. The rubber hand and the concealed real hand were then stroked simultaneously with a paintbrush while the volunteer's brain was scanned using Magnetic Resonance Imaging. Within just 11 seconds the volunteers had started to accept the rubber hand as their own and when later asked to point toward their right hand, most pointed toward the rubber hand in preference to the real one, indicating how the brain had readjusted.

The scientists discovered that one particular area of the brain – the premotor cortex – recognizes the body by accepting information from three different senses: vision; touch; and proprioception (position sense). If there are inconsistencies in terms of the messages received, however, the brain "believes" visual information because it is the strongest sense out of the three. Leading researcher Dr Henrik Ehrrson said: "This study shows that the brain distinguishes the self from the non-self by comparing information from the different senses. You could argue that the bodily self is an illusion being constructed in the brain."

In more serious cases, phantom limb expresses itself as an excruciating pain, variously described as burning, cramping or shooting. Phantom limb pain is thought to be caused by damaged nerve endings. Their subsequent erroneous regrowth leads to the abnormal and painful discharge of neurons in the stump and sometimes changes the way that nerves from the amputated limb connect to neurons within the spinal cord. There is also evidence to support a theory of altered nervous activity within the brain as a result of the loss of sensory input from the amputated limb. One of the treatments for phantom limb pain involves repeated touching of the skin over the stump in order to improve sensory discrimination there. This method has proved effective, possibly because the touching replaces some of the sensory input to the brain that was lost following amputation.

Although phantom limb pain is a physical ailment, in 1996 Dr Vilayanur Ramachandran from the University of California conducted a series of experiments involving intricate mind games. People who suffered from phantom limb pain in the area of an amputated arm were asked to place their arms inside a mirror box so that they saw their remaining arm mirror-reversed to look like their amputated one. When they moved their remaining arm in the box, they were tricked into thinking they were moving their amputated one, and their pain was reduced. Six of the ten trial patients instantly felt their phantom limbs move while a few were able to shift their phantom limbs out of painfully awkward positions. One patient even managed to correct his body image so that his phantom limb eventually shrank away to nothing.

In another pioneering treatment, patients are trained to imagine their amputated arms moving in relation to a moving arm on a screen in front of them. This too has met with some success and is part of a shift in emphasis in the method of treating phantom limb pain by moving away from the site of damage – the stump – and concentrating instead on the centre of pain processing – the brain.

For all the difficulties and discomfort associated with phantom limb, it does have one benefit. Because of the increased feeling that sufferers have, it is easier for them to learn to use certain types of prosthetic device if they are experiencing phantom limb sensation.

Cellular Memory

The notion that a transplant patient can inherit certain personality traits from the donor has long been a mainstay of thriller plots in literature. For example, *Les Mains d'Orlac* by French writer Maurice Renard tells the story of a concert pianist who loses his hands in an accident and, in a transplant operation, is given the hands of a murderer. As a result the pianist suddenly develops an urge to kill. The vast majority of scientists have always dismissed the notion of inheriting memory from donor organs as pure fantasy, but a growing number of experts are beginning to come round to the possibility, persuaded in no small part by some remarkable test cases in which organ transplant patients have reported experiencing marked changes in their food preferences, music tastes and even sexual cravings.

American psychologist Paul Pearsall, himself the receiver of a bone marrow transplant, conducted a series of interviews with heart or heart-lung transplant recipients, their families and friends and the donors' families and friends. The following were among the cases he reported:

- A 29-year-old lesbian and fast food addict received a heart from a young woman who was a vegetarian and also man-mad. After the operation the recipient claimed that meat made her sick and that she was no longer attracted to women, she eventually became engaged to a man.

- A seven-month-old boy received a heart from a 16-month-old boy who had drowned. The donor suffered from a mild form of cerebral palsy, principally on the left side. The recipient, who displayed no such symptoms before the transplant, subsequently developed the same shaking and stiffness on the left side.

- A heart transplant recipient was amazed by his newfound love of classical music. He later learned that the donor was an accomplished violinist who had died in a drive-by shooting.

- A 47-year-old man received a heart from a 14-year-old girl who suffered from an eating disorder. Following the transplant, he displayed childlike exuberance, a little girl's giggle and a tendency to feel nauseous after eating.

Memory is generally considered to be a function solely of the brain, but as a result of his findings, Pearsall suggests that cells of living tissue also have the capacity to remember. Dr Candace Pert, a pharmacologist and professor at Georgetown University, Washington, DC, also believes that the mind is not just in the brain, but throughout the human body. "The mind and body communicate with each other through chemicals known as peptides," she says. "These peptides are found in the brain as well as in the stomach, muscles and all of our major organs. A psychosomatic network extends into the body, all the way out along pathways to internal organs and the very surface of our skin." She believes that memory can be accessed anywhere in this network. "A memory associated with food may be linked to the pancreas or liver," she adds, "and such associations can be transplanted from one person to another."

Another extraordinary case that appears to support the theory of cellular memory was that of an eight-year-old girl who received the heart of a ten-year-old murder victim. After the transplant, the recipient experienced terrifying nightmares of a man murdering her donor. A psychiatrist was so impressed by the lucidity of these images that the police were informed. According to the psychiatrist: "Using the description from the little girl, they found the murderer. The time, weapon, place, clothes he wore, what the little girl he killed had said to him… everything the little heart transplant recipient had reported was completely accurate."

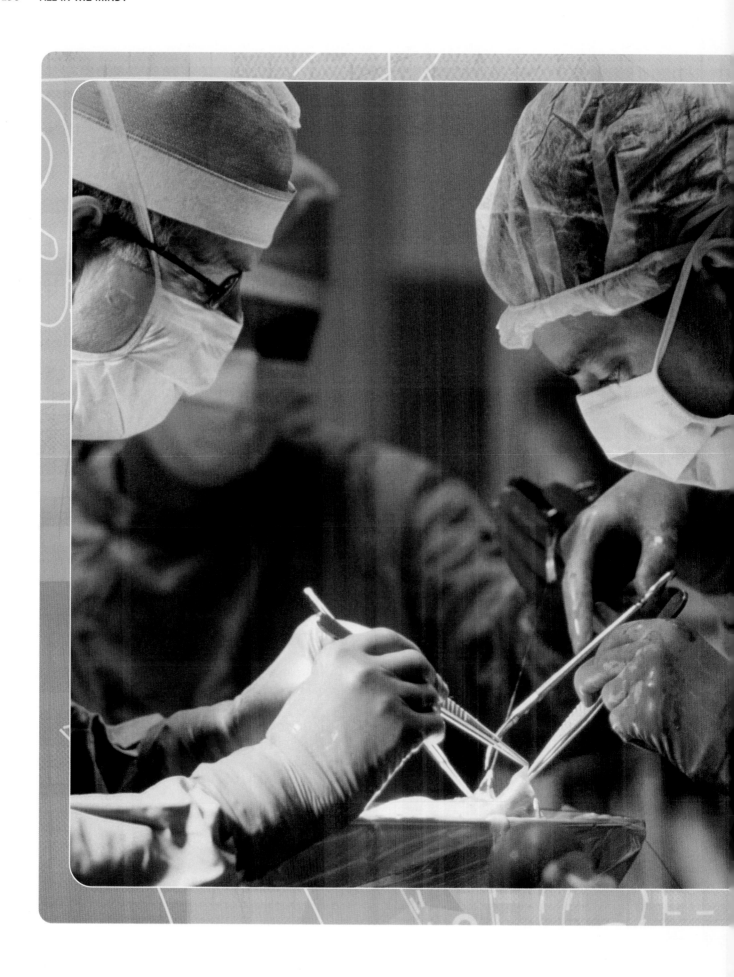

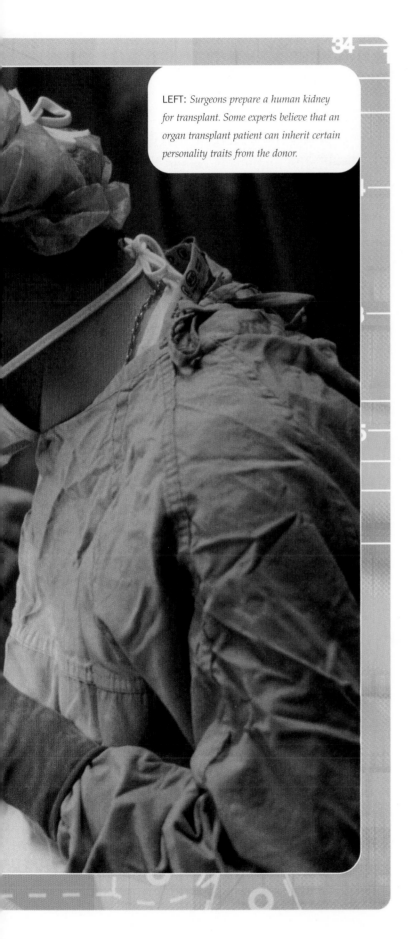

LEFT: *Surgeons prepare a human kidney for transplant. Some experts believe that an organ transplant patient can inherit certain personality traits from the donor.*

An equally bizarre occurrence of apparent memory transfer took place when a young man emerged from transplant surgery and said to his mother: "Everything is copasetic." His mother said that from never having used the word before the operation, he now made it a key part of his vocabulary. It later transpired that the word had been a signal used by the donor and his wife to indicate that they were still friends after an argument. The donor's wife stated that they had argued shortly before the donor's fatal accident and had never made up.

One of the highest profile cases has to be Claire Sylvia, who was stricken by a crippling pulmonary disease in her 40s. To save her life she was given the heart and lungs of an 18-year-old man who had been killed in a motorcycle accident. After recovering from surgery, she began craving beer and chicken nuggets, neither of which she had liked before, and she started having strange and vivid dreams about a young man she didn't recognize. Over the years she came to realize that the young man was the donor and that beer and chicken nuggets were his favourites. She even discovered that he had chicken nuggets in his jacket pocket at the time of his fatal accident. Her dress sense also changed. She eschewed the vibrant reds and oranges that she used to love in favour of cool colours and she began behaving with uncharacteristic aggression and impetuosity.

Although these cases have fuelled the debate regarding the existence of cellular memory, the idea has still found little support from scientists who continue to dismiss these strange phenomena as post-surgery stress or reaction to the immunosuppressive drugs taken to prevent organ rejection following a transplant. The doubters maintain that medication can alter tastes for food and that the enormous feelings of relief at having another chance in life may explain why some patients change their habits and interests. "Most scientists believe psychological experience is stored in the brain," states leading American cardiologist John Schroeder. "The idea that transplanting organs transfers the coding of life experiences is unimaginable." So, who is right? Only time will tell.

The Sixth Sense

Over 2,000 years ago Aristotle recognised the five main senses as sight, hearing, taste, touch and smell. However, there is another sense that is sometimes forgotten. It is called proprioception, which means, literally, "our sense of self". The name was coined by British physiologist Sir Charles Sherrington who called it "our secret sense, our sixth sense". Proprioception was developed by the nervous system to keep track of and control the different parts of the body. It is essential to our awareness of where we are in space and where our arms, legs and other body parts are in relation to one another. It is proprioception that enables us to touch our nose with our eyes closed or to reach up and scratch our head without missing.

The brain receives an extraordinary amount of sensory information and, in order to avoid overloading, it has to prioritise. It has learned to ignore the signals it has come to expect – such as the stretching of our skin when we walk – and deals with these messages in the unconscious parts of the system. Only new or unexpected information reaches the conscious parts of the brain. Every movement we make originates from our brain. When we decide to make a movement, the motor cortex in the brain sends out a command to the relevant muscles to make them move, and within 60 milliseconds a report on the actual movement is sent back from the body's sensors. The brain is constantly scanning signals from our body to be on the lookout for any errors in alignment and co-ordination. For example, even when we are apparently standing still, we always sway slightly from side to side. If our movements are too wide, proprioceptive signals alert the brain, which immediately sends out an order to the muscles to make the necessary adjustments.

Special proprioceptors are located throughout the human body, working in conjunction with the vestibular system, a fluid-filled network within the inner ears that detects the position of our head and helps to keep the body balanced. For example, it is the feedback from proprioception that enables the brain to calculate angles of movement and command our limbs to move precise distances. Proprioceptors in joints, muscles and tendons react to the slightest change in position. They relay the new information back to the brain that monitors these signals, along with those from the eyes, ears and other senses, to control our balance and co-ordinate our movements. This ensures that no one part of the body moves independently from another.

Most of us don't even know we have this "sixth sense", but it is absolutely essential in regard to the movements of the human body. None of us could walk, lift, stretch or dance without our sense of proprioception. Although the most dominant of all the messages our brains receive is the feedback from our eyes, visual information is processed far slower than proprioceptive information. So while dancers traditionally practise in front of mirrors, a dancer relying on information about the state of her body solely from her reflection will be less steady than a dancer listening to her body.

Fortunately, although we sometimes lose our sense of smell or our sense of taste, we very rarely lose our sense of proprioception, but when we do it has disastrous consequences. Ian Waterman from Southampton, UK, is one of only ten people in the world known to have lost his ability to co-ordinate any form of movement unconsciously. One day in May 1971, he cut his finger and the cut became infected, the redness and inflammation soon spreading a little way up his arm. Then he began to suffer alternating hot and cold spells, accompanied by a feeling of general fatigue that forced him to take time off work from his job as a butcher. When he somehow summoned the energy to mow the lawn, he lost control of the motorized lawn mower and could only look on helplessly as it careered away. A week later he fell as he tried to get out of bed and was taken to hospital. By then he was unable to move properly and had no sense of touch and pressure in his hands and feet, although he remained sensitive to temperature and pain.

The viral infection had destroyed the nerves that controlled his proprioception as well as those for feeling light touch. He was deprived of all sensation of touch below the neck. He still had all the nerves for controlling muscle movement, but while his brain was able to send messages to his muscles to make them move, it didn't receive any feedback to say whether the movement had been completed or not. The only feedback he had from the outside world regarding the positioning of his limbs was from his eyes. Consequently, while he could make movements, he couldn't control them. Effectively, this left him paralysed and, to make matters worse, the doctors he consulted struggled to comprehend the full extent of his disability. At first they diagnosed him as having a disorder of the peripheral nerves, from which, they suggested, he would soon recover, but seven months later he was still experiencing difficulty moving. Eventually he was pronounced incurable and was left facing a life confined to a wheelchair.

Whereas a person with full command of the senses can move a finger back and forth easily, without proprioception the brain cannot feel what the finger is doing, and therefore what is normally a simple process requires tremendous thought and planning. The only solution – as Waterman eventually found – is to use vision to compensate for the lost feedback. By watching his body and concentrating on making it move, he finally managed, with a supreme effort, to sit up in bed unaided. "I looked where my legs were before I started. I looked where my arms were. I looked where my body was and then I started to sit up very gradually. And I was so euphoric at having sat up for the first time that I almost fell out the bed because I lost the concentration."

The amount of mental effort required to make the kind of basic movement that most of us take for granted cannot be exaggerated. Waterman has compared his exhausting routine to having to run a marathon every day. His eyes had to be trained to judge the weight and length of objects. As he tried to lift an object, there was no feedback on how hard he had to flex his muscles apart from the clues given to him by his eyes. It took him a whole year to learn to stand on his own, from which point he was able to learn to walk – in the process becoming the first

sufferer of this rare and little understood condition to master the art of walking. He added this and other movements to his repertoire by choreographing each movement step by step.

"I started to develop other movements as well, such as lifting a leg, moving an arm, and then putting some of these together, and gradually, bit by bit, you know if you have enough basic movements you have the building blocks to structure more radical movement, and you can actually move around and be fairly safe, hopefully. I fell over an awful lot in the process, but it was a case of finding these building blocks and then building on that."

The downside with relying solely on vision is that if the lights suddenly go out, he crumples in a heap on the floor, unable to move until they come back on.

Although he has never recovered his proprioception, Ian Waterman was able, by teaching himself visually, to leave hospital after several years of rehabilitation and to resume a life of sorts. He not only learned to walk, but also to care for himself and even drive a car, achieving the last-named function by using vision to estimate his movement, speed and direction. He was also eventually able to hold down a job and to get married. So successful has he been in overcoming seemingly insurmountable odds that, while his movements may look slightly mechanical, few who meet him would suspect that there is anything amiss unless something unexpected happens and he is thrown off balance. But, he conceded recently, "It's been a huge mental drain on me and still takes an awful lot of cognitive energy to maintain my movements."

Ian Waterman's case has allowed scientists to learn a great deal more about proprioception. They were surprised at how accurately he can estimate the weight of objects that he lifts. It is generally thought that people judge properties such as weight and length by using feedback from the stretch of their tendons and muscles. With no feedback available in Waterman's case, he has to use vision to see how his body reacts to a set movement when he picks something up. The faster and higher his body moves,

the lighter the object must be. Indeed, he has become so sensitive visually with regard to how his body reacts that he can detect weight differences of as little as ten per cent between comparable objects. However, with his eyes shut he can only identify weight differences of around 50 per cent.

Another who had to teach herself how to move again after losing her sense of proprioception was the celebrated American dance choreographer Agnes de Mille. One day, in May 1975, she was signing a contract when she discovered to her horror that her hand would not work. Although she felt no pain, she had suffered a stroke that had emptied the right side of her body of all feeling and control. A scan showed that the stroke had affected the thalamus, the area of the brain that receives, processes, and forwards messages from the body's senses. Her sense of proprioception had gone.

Yet, despite being nearly 70, she bravely fought the continuing paralysis in her right side, even overcoming a heart attack and several more small strokes. Like Ian Waterman, she countered the loss of proprioception with visual awareness. Although she had not been expected to survive the initial stroke, she lived for another 18 years and even returned to the stage, directing a production from a wheelchair. In 1988, she received a standing ovation, both for her artistry and her courage.

Out-of-Body Experience

Improvements in medicine have meant that more people than ever are being brought back from the brink of death. Consequently, an increasing number of survivors have reported undergoing "near death experiences" while unconscious during surgery, among them Hollywood actress Sharon Stone who revealed that she had a "white light experience" during a brain haemorrhage scare in 2001. "Near death experiences" seem to take several forms – the appearance of God-like figures in bright light, the sense of moving along a tunnel, a feeling that all ties have been cut, seeing your whole life pass before you in an instant. By far the most common, however, is the "out-of-body experience", the sensation that you are no longer in your body, but are still able to perceive it from the outside.

LEFT *Hollywood actress Sharon Stone said that she had a "near death experience" during her brain haemorrhage scare in 2001.*

While serving as a volunteer ambulanceman in Italy during the First World War, American author Ernest Hemingway was wounded by a burst of shrapnel. As he lay injured and semi-conscious, waiting for medical assistance, he underwent a strange experience. He later described it as "my soul, or something, coming right out of my body, like you'd pull a silk handkerchief out of a pocket by one corner. It flew around and then came back and went in again, and I wasn't dead any more."

At the age of 36, actress Jane Seymour contracted a severe case of flu, for which she was given an injection of penicillin. However, she suffered an allergic reaction, prompting a curious sensation.

"I literally left my body. I had this feeling that I could see myself on the bed, with people grouped around me. I remember them all trying to resuscitate me. I was above them, in the corner of the room looking down. I saw people putting needles in me, trying to hold me down, doing things. I remember my whole life flashing before my eyes, but I wasn't thinking about winning Emmys or anything like that. The only thing I cared about was that I wanted to live because I did not want anyone else looking after my children. I was floating up there thinking, 'No, I don't want to die. I'm not ready to leave my kids.' Although I believe that I 'died' for about 30 seconds, I can remember pleading with the doctor to bring me back. I was determined I wasn't going to die." Then suddenly she found herself back in her body.

Such experiences are not the sole preserve of actors and authors, however. Confined to his bed with severe food poisoning, Dr Auckland Geddes recounted his bizarre story in a 1937 paper to the Royal Medical Society of Edinburgh.

"I suddenly realised that my consciousness was separating from another consciousness, which was also 'me'. Gradually I realized that

I could see not only my body and the bed in which it was, but everything in the whole house and garden, and then I realised that I was seeing not only things at home, but in London as well, in fact wherever my attention was directed. I was free in a time dimension of space." He then recalled how his spirit had re-entered his body. **"I saw my doctor leave his own patients at his surgery and hurry over to my house, and I heard him say: 'He is nearly gone.' I heard him quite clearly speaking to me on the bed, but I was not in touch with my own body and I could not answer him."**

Although not every out-of-body experience (OBE) has identical features, there are a number of common traits. They invariably occur during sleep, on the verge of sleep or while losing consciousness. Often the sleep is not particularly deep, perhaps due to illness, emotional stress or exhaustion. The person then "wakes up" and in many cases experiences a sense of physical paralysis. People remember trying to move limbs during OBEs, but to no avail. Only the eyes seem to function properly. Those who were undergoing life-saving surgery at the time frequently remember looking down at their bodies from above and seeing doctors trying to

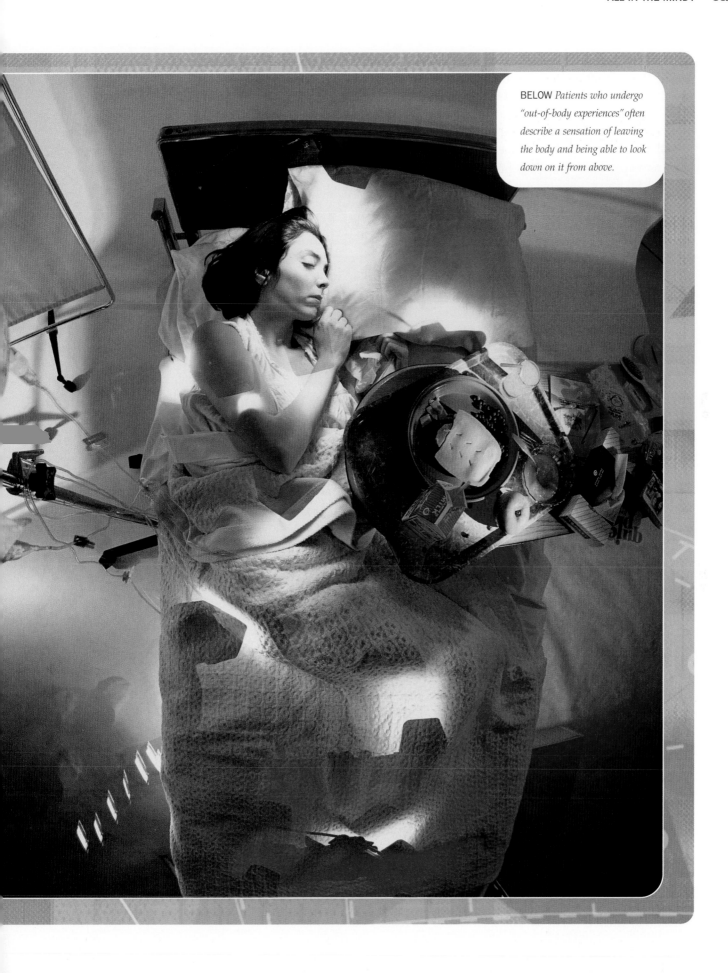

BELOW *Patients who undergo "out-of-body experiences" often describe a sensation of leaving the body and being able to look down on it from above.*

perform resuscitation. The OBE usually lasts for no longer than a minute and is sometimes ended by a fear of getting too far away from the body. Suddenly there is a feeling of being pulled back into the body and the experience is over.

OBEs are often linked to escape from extreme pain and trauma. In his book *The Twenty-Fifth Man*, US prisoner Ed Morrell wrote how he escaped the agony of torture at the Arizona State Penitentiary by floating away from his body. The torture usually involved a specially devised straitjacket on to which sadistic guards poured water to make it shrink. "One seldom survived this treatment for long," wrote Morrell. "A victim being slowly squeezed to death by a boa constrictor can alone appreciate the suffering and anguish of that awful torment." After half an hour of this cruelty, Morrell said his body suddenly felt strangely still. He saw lights flashing before his eyes and found himself separating from his body. The next thing he knew, he was drifting beyond the prison walls and out into the countryside.

On subsequent OBEs he travelled further afield, to distant countries, to outer space and out to sea where he witnessed a shipwreck that he subsequently learned really had taken place. He claimed to have seen people whom he would meet later in life, including the woman he would one day marry. From foreign cities he brought back information to which he had no access through earthly channels and all the while he appeared to be sleeping peacefully, much to the annoyance of his tormentors who tried to invent even more dastardly devices to break him. Toward the end of his sentence the tortures stopped and he found that he had lost the ability to leave his body – perhaps because the OBEs had served their purpose. American author Jack London was so fascinated by the account that he based a novel, *The Star Rover*, on Morrell's story.

Not all OBEs are accompanied by sleep. Susan Blackmore, a leading authority on OBEs, experienced her own while a student at Oxford University in the 1970s. She describes her state of mind at the time as "fairly peculiar" and recounts how she went down a tunnel of trees toward a light before floating on the ceiling and observing her body below. She then floated out of the building and across the Atlantic to New York. After hovering around New York, she drifted back to her room in Oxford where she shrank in size and entered her body's toes.

Eager to sample their own OBE, some people have tried to induce one – usually either by attempting to fall asleep without losing consciousness, taking drugs, or subjecting themselves to stimulation of the brain. In 2002 neurologists at Geneva University Hospital in Switzerland used electrodes to stimulate the brain of a female patient they were treating for epilepsy. They discovered that stimulating one particular spot of the brain – the angular gyrus in the right cortex – repeatedly triggered out-of-body experiences. At first, the stimulations caused the woman to feel that she was "sinking" into the bed or falling from a great height, but when the current was increased, she reported leaving her body. She told the doctors that she could see herself lying in bed from above as she floated near the ceiling. When asked to watch her real legs while the current was passed through the electrodes attached to her head, the patient replied that her legs were becoming shorter. If bent, her legs appeared to be moving rapidly toward her face, causing her to take evasive action. When she was then asked to look at her outstretched arms, she said that the left arm appeared shortened, but the right arm was unaffected. If both of her arms were bent at the elbow, she believed her left lower arm and hand were moving toward her face. The Swiss doctors concluded that the angular gyrus plays a key role in matching up visual information and the brain's touch and balance representation of the body. When the two become dissociated, an out-of-body experience may result. The researchers also noted that the OBE tends to disappear when the person attempts to inspect the illusory body or body part.

The evidence of the Swiss doctors supports the view that at least some OBEs are caused by an unusual brain state in which one's body perception and sense of reality are altered. How much faith can we place in OBEs though? While it is impossible to disprove their existence, as yet there has there been no firm proof that anyone has ever actually left their body during these episodes. Ed Morrell's tale of witnessing a real-life shipwreck may sound convincing, but like other similar claims, its veracity has not been tested to the satisfaction of the scientific world. Until such evidence is forthcoming, many scientists believe that out-of-body experiences are simply vivid dreams, hallucinations or the product of hysteria, brought about by a particularly traumatic event.

The God Spot

Rudi Affolter and Gwen Tighe have both experienced powerful religious visions. He is an atheist; she is a Christian. He thought he had died and gone to hell as a punishment for not being a devout Christian; she thought she was the Virgin Mary, her husband was Joseph and their son Charlie was the baby Jesus. Both Rudi and Gwen have temporal lobe epilepsy.

With temporal lobe epilepsy, abnormal electrical discharges in the brain are confined to one temporal lobe, an area on the lower side of each half of the cerebrum, the main mass of the brain. The temporal lobes are concerned with such functions as smell, taste, hearing, visual associations, and some aspects of memory. Thus any unusual activity there – brought on perhaps by a head injury, an abscess or a stroke – may result in a malfunction of any of these senses. Typical symptoms of temporal lobe epilepsy include dream-like states, hallucinations, and a sense of *déjà vu*. Now research suggests that the temporal lobes are also connected to profound religious visions.

For some years it had been suspected that St Paul, Joan of Arc, Mother Teresa and Mohammed might all have suffered from temporal lobe epilepsy, and that the "sacred disease", as it became known, was somehow linked to their religious experiences. Intrigued by the theory, in 1997 San Diego neuroscientist Dr Vilayanur Ramachandran decided to carry out his own experiments to compare the brains of three groups of people – those with temporal lobe epilepsy, those without the disorder, and those who said they were intensely religious. He examined his subjects' changes in skin resistance, using electrical monitors to measure how they reacted when they looked at 40 different images – violent words, sexual words, simple words and words related to religion. He discovered that when the epileptics were shown any type of spiritual imagery, their bodies produced a dramatic change in their skin resistance, similar to that of the deeply religious group.

ABOVE *The temporal lobes are situated on the lower side of each half of the cerebrum, the main mass of the brain.*

Dr Ramachandran reported: "We found that every time the group with epilepsy looked at religious words like 'God', they'd get a huge galvanic skin response. There may be dedicated neural machinery in the temporal lobes concerned with religion." He concluded that epileptic seizures cause damage to some of the pathways that connect the area of the brain that deals with sensory information to the one that attaches emotional significance to such information. Consequently, these patients can perceive an unusual depth of spiritual meaning in every object and event. He claimed that, following seizures, about 25 per cent of epileptics report deeply moving spiritual experiences, ranging from a feeling of divine presence to a sense of direct communication with God. Although none of the participants suffered actual seizures during the experiment, it was the first piece of clinical evidence to reveal that the body's response to religious symbols was linked to the temporal lobes of the brain. This area has since been dubbed the "God spot".

Evolutionary scientists have suggested that belief in God may be built into the human brain's complex electrical circuitry as a Darwinian adaptation to

LEFT *Ellen G White, founder of the Seventh-day Adventist Church. Could her powerful religious visions have been caused by temporal lobe epilepsy?*

encourage co-operation between individuals. "It may have evolved to impose order and stability on society," said Dr Ramachandran.

Confirmation of the probable link between temporal lobe epilepsy and profound religious experiences has renewed interest in figures of the past. In 2001 an article in *New Scientist* magazine claimed that the prophet Ezekiel may have suffered from temporal lobe epilepsy, contending that he exhibited many of the tell-tale signs – fainting spells, temporary bouts of speechlessness, delusions, and excessive religiousness. More controversially, American neurologist Professor Gregory Holmes believes that Ellen G White, founder of the Seventh-day Adventist Church, also suffered from temporal lobe epilepsy. During her life she had hundreds of dramatic religious visions that helped convince her followers that she was spiritually special. According to Professor Holmes, however, there may be a simpler explanation for her visions. He learned that when she was nine, she suffered a heavy blow to the head, which left her semi-conscious for several weeks. She was never able to return to school and the accident changed her personality completely. Suddenly she became highly religious and moralistic and, for the first time in her life, she began to have powerful religious visions. Professor Holmes is convinced that the blow to the head caused her to develop temporal lobe epilepsy. "Her whole clinical course to me suggested the high probability that she had temporal lobe epilepsy. This would indicate to me that the spiritual visions she was having would not be genuine, but would be due to the seizures." Not surprisingly, the Seventh-day Adventist movement strongly disputes the assertion.

Whether or not religious figures from history did have the disorder is destined to remain a mystery, but the recent research provides another fascinating insight into the intricate workings of the brain and perhaps goes some way to explaining why some people have such intense religious beliefs.

The Placebo Effect

In 2004, researchers from the University of Michigan and Princeton University placed a number of volunteers inside MRI machines and applied either electric shocks or heat to the volunteers' arms. The pain activated all the expected neural pathways. Then researchers smeared on a cream, which they said would block the pain. In fact it was nothing more than a regular skin lotion that would not provide any measure of pain relief. Yet when the volunteers were again exposed to pain, they reported considerably less sensation – statements borne out by pain circuits in the brain, which showed that they really did feel better. These were the same regions of the brain that respond to painkilling medication. However, when the researchers spread the same cream on again, this time revealing its true, ineffective nature, there was no pain relief at all.

The experiment was a classic demonstration of the placebo effect. A placebo (Latin for "I shall please") is a medication or other kind of treatment that seems therapeutic, but in reality is inert and ineffective. Popular placebos include sugar pills and starch pills. The person administering the treatment knows that it contains no remedial properties, yet the patient believes it to be beneficial and demonstrates a detectable improvement in health as a result of taking the placebo. This is the placebo effect – when a patient's symptoms can be alleviated by an ineffective treatment simply because the individual expects or believes that it will work. The inference is

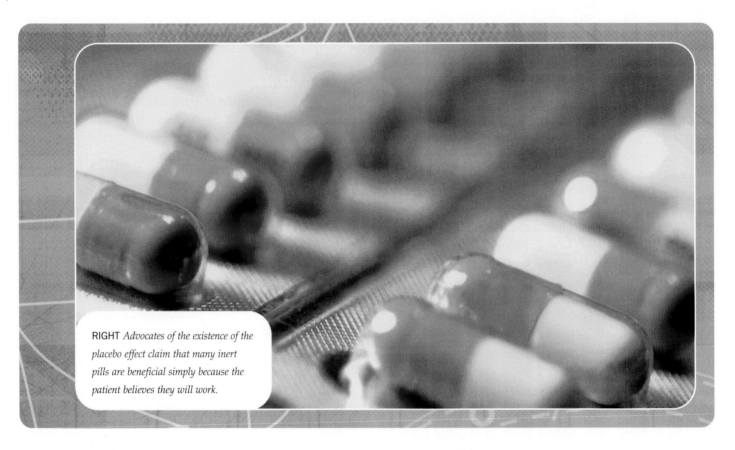

RIGHT *Advocates of the existence of the placebo effect claim that many inert pills are beneficial simply because the patient believes they will work.*

that curing certain diseases relies more on a mental attitude than expensive drugs. Understandably, it is a concept that does not exactly find much favour with the pharmaceutical industry.

A similar study was carried out by researchers from the University of British Columbia in Canada on a group of patients suffering from Parkinson's disease. Positron Emission Tomography (PET) was used to study the brains of the patients while they received treatment. Some were injected with an active drug; the others, although they didn't know it, were given a harmless placebo. The team measured the amount of dopamine released by the brain's damaged neurons, which indicates the effectiveness of drug treatment on the disease, and found that those who received placebo injections exhibited considerable dopamine releases. In fact, the placebo effect produced the same results as pharmaceuticals.

In another test, performed on ten patients scheduled for an operation to relieve the arthritic pain in their knees, half of them underwent placebo surgery – in other words there was no recognized surgical procedure at all. Instead the surgeon simply stabbed the patients' knees three times with a scalpel to create the expected incisions and scars. Six months later, none of the ten patients had any idea which of them had been duped, but all reported significantly less pain in the affected area.

The very existence of the placebo effect has been the subject of medical debate for years. The phrase was coined in 1955 by Harvard anaesthesiologist Dr Henry Beecher who calculated that around one-third of patients he studied showed a noticeable improvement due to placebos. In certain areas – such as pain, depression, heart ailments and gastric ulcers – placebos have subsequently been claimed to be effective in up to 60 per cent of patients. Irving Kirsch, a psychologist at the University of Connecticut, goes as far as suggesting that the effectiveness of Prozac and similar antidepressant drugs may be attributed almost entirely to the placebo effect. After analysing 19 clinical trials of antidepressants, he concluded that the expectation of improvement, as opposed to adjustments in brain chemistry, accounted for 75 per cent of the drugs' effectiveness. "The critical factor," he says, "is our beliefs about what's going to happen to us. You don't have to rely on drugs to see profound transformation."

With belief being all-important, studies of placebos are usually conducted in "double-blind" fashion – not only are the patients unaware that they are receiving a placebo, the doctors don't know either. Of course, some patients do not react favourably to the news that the medication that has apparently relieved their symptoms is nothing more than a dummy pill. It is telling them that there is nothing really wrong with them and that it is all in their mind.

Trials have repeatedly shown that just thinking a medicine will make you feel better will actually have the desired effect – even if it is fake. That is one reason why doctors talk up the benefits of a drug as they write the prescription. Dr Kenneth Casey, a professor at the University of Michigan, says the results of placebo trials should send out a message to doctors. "If you're providing a treatment to a patient, it's important that you realistically provide them with the expectation that it would work, so you enhance the effect. If you gave them a drug or any kind of treatment with the attitude, either explicit or implicit, that this might not be effective, it would be much less likely to be effective."

In a study of asthmatics, researchers found that they could produce dilation of the airways just by telling people they were inhaling a bronchiodilator, even when they weren't. Also, patients suffering pain after having their wisdom teeth extracted derived as much relief from a fake application of ultrasound as from a real one, so long as both patient and therapist thought the machine was switched on. Equally remarkably, doctors managed to remove warts by painting them with a brightly coloured, inert dye and promising the patients that the warts would be gone when the colour wore off.

Around 1960, Leonard Cobb, a Seattle cardiologist, conducted a trial on a medical procedure commonly used at the time to treat angina, whereby doctors made small incisions in the chest and tied knots in two arteries in an attempt to increase the blood flow to the heart. Ninety per cent of patients reported an improvement in their condition. However, when Cobb performed placebo surgery, in which he made incisions but did not tie off the arteries, the sham operations proved just as successful. As a result, the technique, known as internal mammary ligation, was soon abandoned.

However, the placebo effect would appear to be more than purely psychological. As the tests at Michigan and British Columbia indicated, a patient's hopes about a treatment can produce a significant biochemical effect. Sensory experience and thoughts can affect neurochemistry, so it follows that a person's optimistic attitude and beliefs may play an important part in their recovery from injury or illness. One theory is that the placebo effect serves to stimulate the release of endorphins, the chemicals in the body that relieve pain.

It is also thought in some quarters that the component parts of medical treatment – the sympathy, the care, the interest, etc – can trigger physical reactions in the body that promote healing. US psychiatrist Dr Walter A Brown told the *New York Times Magazine*: "There is certainly data that suggests that just being in the healing situation accomplishes something. Depressed patients who are merely put on a waiting list for treatment do not do as well as those given placebos."

Placebos do appear particularly effective in the treatment of patients with anxiety or depression. Back in 1965, researchers gave 15 outpatients at an American psychiatric clinic an inert pill. Furthermore, they openly identified it as such, telling the patients beforehand that it was a sugar pill with no medicine in it whatsoever. A week later, 14 of the patients reported an improvement in their condition (the other one dropped out after her husband mocked the idea).

Doubters dismiss the so-called placebo effect as little more than coincidence, maintaining that any improvement in a patient's condition is merely a result of an illness or injury following its natural course. Even long-term ailments, particularly aches and pains, can suddenly vanish overnight without any form of actual treatment. However, studies often show that people who are given no treatment at all fail to demonstrate the same improvement as those given placebos.

The power of the placebo effect has created an ethical dilemma as to whether doctors should knowingly deceive patients, even though it may be for their own good. Some "alternative" practitioners take the view, perhaps understandably, that as long as the treatment is effective, who cares if it's a placebo?

Scientists have yet to come up with a wholly plausible explanation for the placebo effect. It may be physical, it may be psychological, or it may even be a combination of both. According to some, the whole idea may just be a myth. Whatever the reality, the argument surrounding it, however, it shows no sign of abating.

Anarchic Hand Syndrome

It sounds like something out of a horror movie. A woman wakes up in the middle of the night with a start to find a hand gripped around her neck. With her right hand, she desperately tries to drag it off until she realises that the attacker is her own left hand. But this is no screenwriter's fantasy: this an example, albeit an extreme one, of a rare and disturbing disorder known as anarchic hand syndrome, where, quite literally, the left hand doesn't know what the right hand is doing, or *vice-versa*.

Also known as alien hand or Dr Strangelove syndrome (after the German scientist with the unruly limb in the 1960s movie starring Peter Sellers), anarchic hand syndrome is a condition where one of the sufferer's hands seems to take on a life of its own. Although sufferers can feel normal sensation in the hand, they become convinced that it is not part of their body and that they have no control over its movements. Anarchic hands can also perform complicated acts, such as undoing buttons or removing clothing. Sometimes sufferers are not aware of what the rogue hand is doing until it is brought to their attention and then they often become angry with it, fighting or punishing it in an attempt to control it and claiming that it is "possessed" by some malevolent spirit. Just as Dr Strangelove constantly struggled to stop his arm doing a Nazi salute, people with anarchic hand syndrome can be seen slapping and grabbing their errant hand to stop it misbehaving. Patients with the phenomenon behave as if they have two streams of consciousness, or two separate wills, which compete with each other. This may lead to one hand "arguing" with the other, for example over choosing which television channel to watch.

Professor Sergio Dalla Sala, an Italian neuropsychologist based at Aberdeen University in Scotland who made a special study of the disorder, said: "One patient picked up fish bones from a meal and put them in her mouth. Then she was very embarrassed and tried to take them out, and her two hands got into a fight. I have also seen patients who cannot avoid grabbing a very hot cup with the anarchic hand. They say, 'No, no, it's too hot', but they still grab it." Another patient tied one hand behind her back and adamantly refused to release it. "She said it is mad," added Professor Della Sala.

More dramatically, a patient was happily driving home when the anarchic hand suddenly seized the steering wheel and nearly caused an accident. In other cases, people have been thwarted in the simple of task of writing their own name because the reckless hand keeps pushing the controlled hand away from the writing paper.

There have only ever been about 40 documented cases of anarchic hand syndrome. It is caused typically by trauma to the brain, after brain surgery, or after a stroke or an infection of the brain. Different types of brain injuries cause different subtypes of anarchic hand syndrome. In a right-handed person, an injury to the *corpus callosum* (the area of the brain that connects the two cerebral hemispheres, the two halves of the brain) can give rise to uncontrollable movements of the left hand. However, an injury to the frontal lobe of the brain can trigger grasping behaviour and compulsive manipulation in the dominant right hand. An injury to the brain's cerebral cortex, which controls conscious thought, sensation and movement, can bring about involuntary movements of either hand. More complex hand movements, such as the unbuttoning or tearing of clothes, are usually associated with brain tumours, aneurysms or strokes.

It is thought that anarchic hand syndrome occurs as a result of disconnection between different parts of the brain that control bodily movement. For example, the two hemispheres of the brain are sometimes surgically separated in order to relieve the symptoms of extreme cases of epilepsy. Consequently different regions of the brain are able to command bodily movements without apparently being aware of what the other brain regions are doing. In effect, the mind is divided.

Professor Della Sala thinks that the syndrome can help to explain the basis of free will. Neurologists believe that there are regions in the human brain that regulate actions that are driven by inner will, while inhibiting actions are triggered by the environment. When these areas are damaged, the person is left at the mercy of environmental triggers and their actions do not match their will.

"It raises the question, how free is our free will? It seems to demonstrate that self-ownership of actions can be separated from awareness of actions. Anarchic hand patients seem to be aware of the actions of the anarchic hand but they disown them. The patients are aware of the bizarre and potentially hazardous behaviour of their hand, but have great difficulty inhibiting it. They often refer to the feeling that one of their hands behaves as if it has its own will, but never deny that this capricious hand is part of their own body."

Sadly there is currently no treatment for anarchic hand syndrome. All patients can do to control the problem is to keep the hand occupied by holding something in it.

ABOVE *Peter Sellers as the fictional German scientist Dr Strangelove, the most famous sufferer of anarchic hand syndrome.*

7 UNUSUAL ALLERGIES

An allergy occurs when the body's immune system reacts inappropriately to certain substances. The allergic reaction occurs only on the second or subsequent exposures to the offending agent, after the first contact has sensitized the immune system. Typically, although a substance will cause an adverse reaction in some people, the majority of the population will experience no symptoms whatsoever. Hay fever is a prime example. While most of us enjoy being outdoors in the summer, hay fever sufferers are subjected to eye irritation, runny noses and bouts of sneezing. For many, the condition is so bad that they have to stay indoors rather than risk contact with the offending pollens. Unfortunately flower, grass and tree pollens are among the most common allergens, along with house dust, feathers, animal fur, bee and wasp stings, milk, eggs, shellfish, strawberries and nuts – all of us probably know someone who is allergic to at least one item on this list. Because these allergies are well known, they are accepted by society as part and parcel of living in the twenty-first century. However, other allergies are so unusual that they have yet to be scientifically recognized, consequently the sufferers are frequently ridiculed and their symptoms are considered to be psychological rather than medical.

Allergic to Sex

British model Emma Jones was alarmed when she came out in a painful rash, suffered terrible headaches and was left gasping for breath after enjoying sex with the new man in her life. She attributed it to the stress caused by the break-up of her first marriage, only to be told by doctors that, like many women, she was allergic to the latex in condoms. On this advice, she and her French husband Stephen tried sex without a condom but, still she suffered the same symptoms. This time the diagnosis was more devastating – she had become allergic to semen.

Emma's problems appeared to start with the end of her first marriage in 2000. It was a highly traumatic period in her life and her body seemed to react by developing allergies. Previously she had been able to eat whatever she liked, but suddenly she became allergic to wheat and gluten. Her weight ballooned by 28lbs (12.7kg), although the weight dropped off again as soon as she stopped eating wheat and gluten. Then hair began to fall out from the top of her scalp and she had to cover it up with a fake ponytail. She only discovered that her allergies extended to sex when she met Stephen.

Emma's first husband had been her first sexual partner and during their 13 years together they had never used condoms. When she met Stephen, however, they used a condom during sex and that was when she first noticed the soreness and headaches. At first she didn't link it to sex, but when the same symptoms occurred again, she went to see her doctor. He carried out skin tests and concluded that she was allergic to the latex in condoms, which brought her out in a rash across her stomach and genital area. She also suffered breathing difficulties, similar to an asthma attack, dizziness and headaches.

As her relationship with Stephen developed, they decided to explore alternative methods of

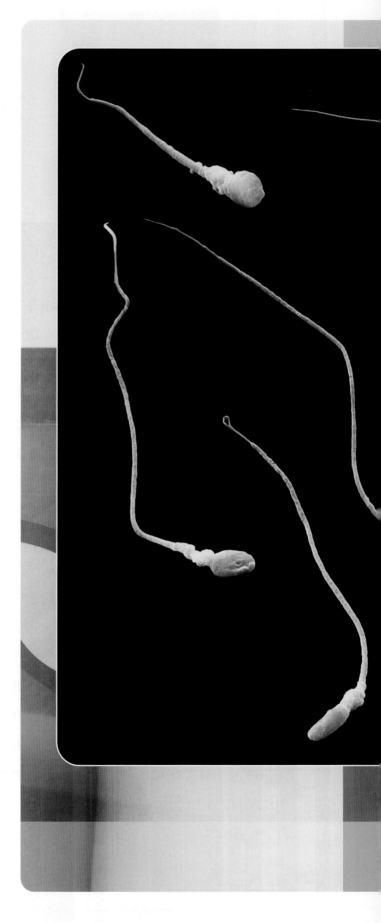

contraception, but she was devastated to find that she still had the same symptoms. And this time, when she went to the doctor, tests revealed that she had become allergic to semen.

"People snigger when they hear about my allergy," Emma told the *Mirror* newspaper, "but it's not a laughing matter. Every time I make love I come out in a painful rash, suffer excruciating headaches, hot flushes and have breathing difficulties. I don't know what it is in the semen that I've grown allergic to, but there's no doubt I am. I've been told to avoid all latex and I always have to read labels very carefully. I can't go anywhere near balloons and I have to be vigilant because I've been warned that the symptoms can worsen and it's possible that I could go into the type of anaphylactic shock that kills people."

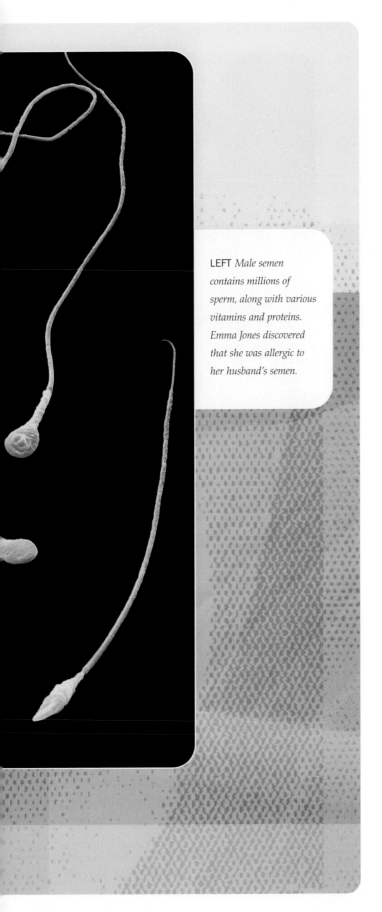

LEFT *Male semen contains millions of sperm, along with various vitamins and proteins. Emma Jones discovered that she was allergic to her husband's semen.*

Although Emma finds that taking an anti-histamine tablet reduces the symptoms, it has been a distressing time for her. Latex allergy has become quite common in recent years – whether it's condoms, latex gloves or balloons – but semen allergy is quite rare. When a man ejaculates, his semen contains millions of sperm plus sugar, vitamins, minerals and proteins to support the sperm on their journey. Occasionally the woman's immune system will recognize one of these components as a foreign substance. When this happens, the body forms antibodies and releases chemicals, resulting in a burning and itching sensation around the vagina.

In extreme cases, it is also possible to suffer a severe systemic allergic reaction, producing swelling of the throat, decreased blood pressure, difficulty breathing, wheezing, fainting, shock and even death. The dangers were highlighted recently after a 25-year-old Romanian woman died because she was allergic to the proteins in her husband's semen. She would choke and feel sick after having sex without a condom and was told by doctors to avoid all contact with human sperm. Nevertheless it appears that the couple still didn't always use a condom, and on the last occasion she suffered an even worse reaction and was rushed to hospital where she died.

It's in his Kiss

Scientists in America have recently discovered something that many of us have known all along – kissing can be a dangerous business. In fact they say that kissing the wrong person can land you in hospital – not as a result of a beating from an angry spouse, but because of something that person may have eaten earlier in the day.

A study in the *New England Journal of Medicine* found that an unexpectedly large number of people with nut allergies developed an itch when kissed by someone who had eaten nuts up to six hours prior to the embrace. In most of the 17 cases investigated, the allergic reaction came on within a minute and was mild – mainly itching and swelling in the area that has been kissed – but five people started to wheeze or flush with light-headedness. And a kiss on the cheek of a three-year-old boy caused a respiratory reaction severe enough to require a trip to the emergency room. What particularly surprised the investigators was the length of time after eating the food that the allergen remained present in the mouth of the kissing partner. Some couples had tried to take precautions, such as brushing teeth and using mouthwash, but these proved ineffective against preventing an allergic reaction in their partner.

In the US alone, serious food allergy puts 30,000 people in hospital each year and kills up to 200 – at least three times as many as die from insect stings. Around half the people with food allergies are sensitive to peanuts or tree nuts, such as walnuts, cashews and almonds. People with a severe allergy to peanuts can even develop low-grade reactions while aboard planes on which the snack has been given to other passengers. Doctors have also treated allergy patients who have reacted to someone eating a peanut in the same room.

Shellfish can produce an equally dangerous reaction. A 20-year-old American woman, who is allergic to shellfish, experienced a severe allergic reaction immediately after kissing her boyfriend.

The boyfriend had eaten several shrimps less than an hour earlier. The couple worked at a seafood restaurant, and she sometimes wore gloves while serving food. She had repeatedly touched shellfish while at work and had consequently experienced a series of mild allergic reactions in the past. It is possible that these minor reactions may have served to "prime" her immune system to produce more antibodies directed at crustacean proteins – a phenomenon that is similar to the reactions people experience to seasonal allergens, such as noxious pollens. On the night of her severe reaction, she reported no distress or symptoms prior to the kiss, but no sooner had they embraced than she suffered a reaction in her lips and skin. Her throat started to

LEFT *Kissing can bring out an allergic reaction – particularly if one of the participants has recently eaten nuts or shellfish.*

swell and she was gripped with abdominal cramps, nausea and wheezing. Haematologist David Steensma says: "It is important to warn susceptible patients that food does not actually have to be eaten to trigger an allergic reaction. Touching the offending food and kissing or touching someone who has recently eaten the food can be enough to cause a major reaction."

Doctors in Naples, Italy, who were called to treat a woman with swollen lips found that she, too, was allergic to her husband's kiss. However, in this instance it was not caused by food but by an antibiotic drug he was taking to combat an infection. They proved their point by giving the husband another dose of the same medicine and getting him to kiss his wife again. Sure enough, 20 minutes later she came out in a rash. The 2002 case is thought to be the first instance of someone suffering an allergy to medicine taken by a partner.

So could these allergies sound the end of the traditional kiss? Certainly not, say experts. US allergy specialist Scott Sicherer promises that "a peck on the cheek is unlikely to cause a severe problem". However, he warns that passionate kisses do increase the likelihood and duration of exposure to another's saliva and could therefore present "some real danger". The best advice, however, is to always tell your partner about your allergies – and to check what he or she has been eating recently before you kiss them.

Multiple Chemical Sensitivity

It goes under many names: Multiple Chemical Sensitivity; idiopathic environmental intolerance; total allergy syndrome; environmental sensitivity; ecologic illness; total immune disorder syndrome; chemical AIDS; or twentieth century disease. Each of these names has specific implications regarding the underlying cause, mechanism or symptoms of the disease. The difficulty in agreeing upon a definition or a name for the condition has been a major obstacle to giving it the scientific recognition that its victims are convinced it warrants.

One thing upon which the various experts are agreed, however, is that it is very much a modern disorder. The popular theory is that it dates from the years after the Second World War when a wide range of new chemical-based products were introduced to the market place, including pesticides, perfumes, paints, glues, solvents, plastics, carpets, shampoo, detergents, medications, soap, caffeine and food additives, to name but a few. These products are now part of our everyday lives; they are in the food we eat, the clothes we wear, the air we breathe. Many such items were not adequately tested for their potential toxicity, with the result that they have triggered unpleasant physical reactions. As far back as the early 1950s, Chicago allergist Theron Randolph noted that people were getting sick from their environment, and within a decade pollution had become a serious health issue. By the 1970s the construction industry was being urged to become more energy efficient and this led to changing the ventilation requirements in new buildings. A combination of the ventilation changes and the increase in volatile chemicals being used in manufacturing materials resulted in what is now known as sick building syndrome, where office workers *en masse* report headaches, feelings of nausea and other allergic reactions.

The symptoms of Multiple Chemical Sensitivity (MCS) are similar to those of traditional allergies, but, because different people react to different

RIGHT *Dangerous levels of pollution over Paris in 2003. Chemical-based products in the air that we breathe can trigger extreme environmental sensitivity.*

LEFT *Experts think that lengthy exposure to such chemicals as those found in pesticides can lead the human body to loose its ability to detoxify.*

products, the list reflects the diversity of the disease. Breathing difficulty, migraines, rashes, dizziness, nausea, fatigue, insomnia, aches and pains, loss of concentration and even amnesia have all been reported by sufferers. Women appear to be more affected than men. It is thought that the testosterone that men produce covers up the symptoms and warning signs until it is too late, although it could be that women are more susceptible because they use more chemical products, such as cosmetics and household cleaners. In one of the most alarming cases, Sheila Rossall, singer with the 1970s pop group Pickettywitch, suddenly developed an extreme sensitivity to all man-made fibres, plastics and processed foods, causing her to swell up and vomit. Because she seemed allergic to everything around her, she tried to live in a virtual bubble and was forced to withdraw to a darkened, air-filtered room in Bristol, UK. Nevertheless her weight dropped to 88lbs (39.9kg) and at one point she became too weak even to lift her head.

So what causes the body to react in the way that it does? Clinical ecologists believe that repeated small exposures or one large exposure to particular chemicals over a period of time can result in the body losing its ability to detoxify. A person with MCS cannot get rid of chemicals in the body because they are entering the body faster than they can be eliminated. The chemicals are then stored in the fatty tissues of the body – the heart, liver and brain. Once the body's ability to process toxins is damaged, the person may be susceptible to chemicals that never previously presented a problem. This susceptibility can manifest itself in a sensitization to products that do not trouble other people and a malfunction of the immune system. In an attempt to define MCS, one scientist has described it as "a chronic condition that is characterized by multiple symptoms, in multiple organs, affecting multiple senses, and triggered by multiple chemicals".

To complicate matters further, there is evidence that MCS and its related diseases are not caused solely by chemicals. Viruses, severe emotional or physical trauma (especially in childhood), liver damage and metabolism disorders have all been held responsible. Some experts are also convinced that the basis for MCS is partly psychological and that the majority of patients also suffer from depression or anxiety. In particular, recent studies at the University of Toronto have linked MCS to panic disorder.

Although the condition is often likened to an allergy, it differs from allergies in one important way. Tests have revealed that when MCS sufferers are in close contact with one of their trigger items, for example a solvent, but are unaware of its presence because researchers have managed to mask its scent, they do not consistently experience symptoms. By contrast, people who suffer from pollen allergies or who are allergic to nuts, for instance, will have an allergic reaction when exposed to their trigger, regardless of whether or not they are aware of its presence. Sensing that there might be a cognitive component to the condition and observing that MCS symptoms are similar to those of panic disorder, the Toronto researchers decided to investigate whether the two were linked. An earlier study had shown that patients with panic disorder had more receptors for a chemical called cholecystokinin, a natural substance made in the human gut and brain. In the gut, it plays a role in digestion; in the brain, it is thought to be related to expressions of anxiety and fear. It is recognized to be a panicogenic agent, meaning that it will induce attacks in patients who suffer from panic disorder. However, it will not provoke an attack in those who do not. To all intents and purposes, it is a test for panic disorder. So, in view of the fact that MCS and panic disorder shared so many characteristics, the researchers decided to see whether the two were also genetically connected.

All humans have two types of cholecystokinin receptors – A and B. Type B comes in 15 different variations, called alleles. Our genetic code determines which alleles we carry. Allele seven has been found to be more prevalent in patients with panic disorder than in the general population. So the Toronto team, led by Dr Karen Binkley, studied a group of 11 patients with MCS, comparing them with 11 people who did not have the condition. Of those with MCS, 41 per cent carried allele seven; in those without, the figure was only nine per cent.

Obviously the test sample was small and considerably more work needs to be carried out before any firm conclusions can be drawn about the psychological aspect to MCS. But Dr Binkley thinks she is heading in the right direction toward discovering more about the underlying causes of this most baffling of disorders. "I think that the distinction between mind and body is really an artificial one," she says. "They function as a whole. You can't view one without the other."

Honeymoon Rhinitis

Bouts of sneezing are brought on by irritation of the nasal passage or upper respiratory tract. The irritation may be caused by inflammation of the tract, which occurs in a number of circumstances – the common cold, influenza and hay fever; inhaling an irritant, such as dust or pepper; or by the presence of mucus. All of these are well-documented, but there is another lesser known instance that can occasionally produce sneezing – a sexually related allergy that has been given the name "honeymoon rhinitis".

From time to time, reports appear in medical journals about men and (to a lesser extent) women who begin sneezing violently immediately prior to sexual intercourse. In some cases it is not even necessary for the sex act to be imminent – simply having erotic thoughts can be enough to bring on sneezing or a constant running of the nose. According to some experts, this is because the lining of the nose is an erectile tissue. Erotic stimuli lead to an increase in the blood supply in the nasal lining, which in turn causes the nose to run. Thus sexual excitement can give you a runny nose or a bout of sneezing.

More alarmingly, sexual arousal can sometimes lead to an asthma attack. Postcoital asthma or "sexercise-induced asthma" is the name given to any asthmatic episode that cannot be ascribed to any cause other than sexual excitement. Such incidents usually occur when one or other of the partners is apprehensive or anxious. The term "postcoital asthma" can be misleading because, like honeymoon rhinitis, it can be brought about by nothing more than intimate contact before the actual sex act. Indeed, it often prevents satisfactory completion of coitus, thereby leading to further anxiety and aggravation of the attack. To prove that the condition is not a form of asthma activated by exercise, a study was conducted in which sufferers were asked to climb two flights of stairs, a movement considered equivalent to the energy expended during sexual intercourse. However, the patients showed no symptoms of rhinitis on these occasions, confirming that it is intense emotional stimuli, rather than physical exertion, that causes the problem.

Writing in the *Journal of the Royal Society of Medicine*, doctors from St George's Hospital, London, detailed another case linking sexual activity and nasal irritation. A man who had taken Viagra to boost his sexual performance ended up in hospital for nearly a week with an unstoppable nosebleed. The man, who was in his late 50s, told doctors that he had engaged in energetic sexual activity in the hours before his first nosebleed. To enhance his performance, he had taken a 50mg dose of Viagra. The doctors had also seen another patient who suffered a two-day nosebleed after taking Viagra. Both men had high blood pressure, a recognized risk factor for heavy nosebleeds, and the doctors theorized that Viagra may act on the nose as well as the penis. Comparing it to honeymoon rhinitis, they suggested the drug could have engorged the veins in the nose, thereby increasing the risk of heavy bleeding.

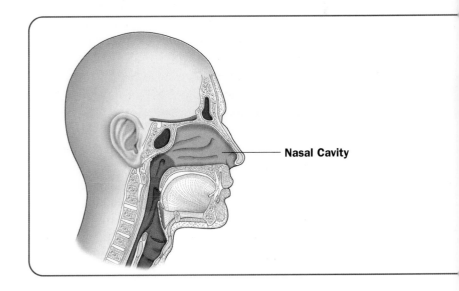

— Nasal Cavity

ABOVE *Sneezing is caused by irritation of the nasal passage. Among the lesser-known irritants is a sexually related condition known as "honeymoon rhinitis".*

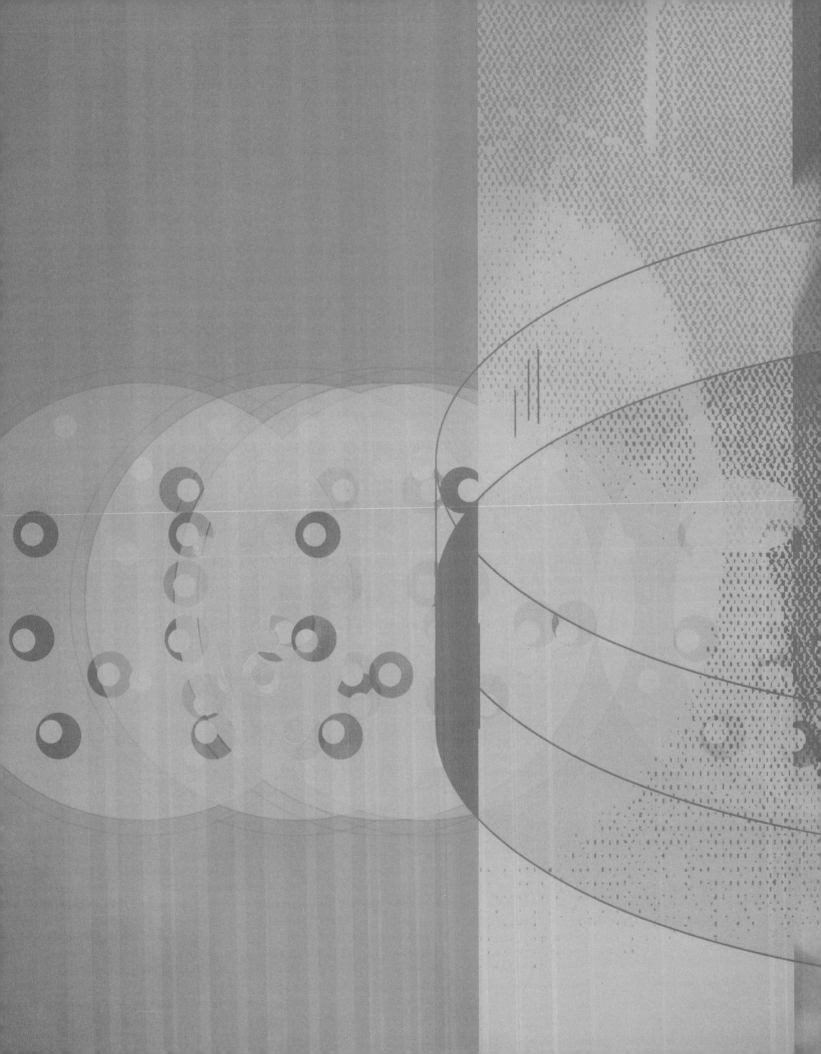

8
SCIENCE IN PROGRESS

Across the globe scientists are constantly working on ways of tackling disease, on improving the quality of our lives and ultimately prolonging them. An indication of the strides that have been made in recent times can be gauged by the fact that during the 1990s the death rate from breast cancer fell by as much as 21 per cent in the UK. Diseases such as smallpox, diphtheria and polio, once so rampant, have been all but eliminated in developed countries. In the early 1950s there were 50 million cases of smallpox a year throughout the world – killing one in every four victims – but by 1980 the World Health Organisation declared it to be extinct. Meanwhile, aided by increased knowledge and tremendously improved technology, the quest continues to combat other deadly diseases. Such life-saving measures are universally welcomed, but other areas of advanced medicine are proving more controversial. The fear is that in trying to enhance our lives, as opposed to actually saving them, scientists are tampering with nature to an unacceptable degree. Nowhere is this argument more keenly expressed than in the debate surrounding the cloning of human embryo. Harnessed correctly, however, science can only be of benefit to the human race so it is our duty to ensure that technology acquires a conscience.

Putting a Smile Back on Children's Faces

Surgeons are using a revolutionary new implant to help re-build the faces of children injured in accidents or born with serious deformities. Around 50 youngsters, including a baby born with a jaw tumour and a 12-year-old girl who had barely been able to open her mouth since birth, are now able to smile again thanks to this life-changing procedure. British and Russian scientists have joined forces to create the honeycomb-like polymer implant, which readily bonds with bone without causing any adverse reaction. The made-to-measure implants are light, tough, flexible and cheap and are seen as providing an excellent alternative to traditional titanium structures.

The PolyHap implant has been developed by teams led by Professor Steve Howdle from Nottingham University and Dr Vladimir Popov from the Institute of Laser and Information Technology in Troitsk, near Moscow. Early in 2005 it was undergoing clinical trials in Moscow where it was being tested on children ranging in age from 18 months to 18 years. Among the patients who have undergone surgery is 12-year-old Kseniya Gordeeva, who suffered jaw damage at birth, as a result of which she struggled to open her mouth. She had to eat through a straw, had difficulty talking and found it almost impossible to clean her teeth. Because of the lack of normal nutrition she was also underweight for her age. During a five-hour operation, the section of damaged bone was removed and a two-inch implant was inserted into her face. Nine days later she was able to open her mouth without so much effort and could eat, laugh and talk to her friends like any other young girl. "If I wanted to get my mouth open before the operation I had to lean my head right back," she said. "Now it is much easier. I can talk like my friends and eat normally. I don't have a favourite food because everything I eat is special." Professor Vitaly Roginsky, one of Russia's leading children's cranio-maxillofacial surgeons, carried out the facial corrective surgery for the trials. He said: "Kseniya has made remarkable progress in a short time. Now she will be able to eat properly and grow into a fine, pretty girl."

He also treated 15-year-old Anara Djantemiroma, who suffered under-development of the jaw as a young girl. This was corrected in a series of operations, the final one involving the insertion of an implant. Professor Roginsky said: "These implants allow us to carry out many more operations than before. They are easier to adjust and reshape and give us much more flexibility in our work."

When a child is assessed for an operation at the St Vladimir Children's Hospital, Moscow, scientists use X-rays and tomography images (an imaging technique that produces a cross-sectional image of an organ or part of the body) to create a three-dimensional plastic cast of the damaged area. These solid models, built by a high-tech process called laser stereolithography, allow surgeons to plan operations with great precision before they even lift a scalpel. Having assessed how much bone needs to be removed, the scientists use stereolithography to make the individual PolyHap implants. The technique, which can be completed in a matter of hours, can be used to make the most intricate shapes, which are then sent to the hospital. The outline of the implant is initially "drawn" by a laser beam that leaves a very fine coating of polymer. This process is repeated hundreds of times until the model is complete. The first Moscow operations were carried out to correct jaw or skull deformities, but the implants can be adapted for any part of the skeleton. The key to the implants' success is the introduction of a mineral-like substance called hydroxyapatite, which makes the polymer tough and "bone-friendly". The collaborating scientists have also found a way to increase porosity, which is important for new bone growth, and clean out toxins from polymers using high-pressure carbon dioxide. Without this process, the implants could cause damaging reactions in the patients.

Dr Popov enthused: "I am convinced polymers will take over from titanium in surgery in the coming years. Now we have found a way to make them stronger, they are ideal for implants. Our technique allows operations to be performed more quickly and efficiently, which is better for the patient and saves time and money for the hospital."

Although the PolyHap implants have produced impressive results, there is a possibility that they might have to be replaced as the child grows and bones develop. With this in mind, the British and Russian teams have started work on a biodegradable version that will slowly dissolve as the repairing bone begins to re-grow. "If we can push the development on to this stage," said Professor Howdle, "it will mean children will only have to undergo one operation rather than several. The benefits from that are obvious."

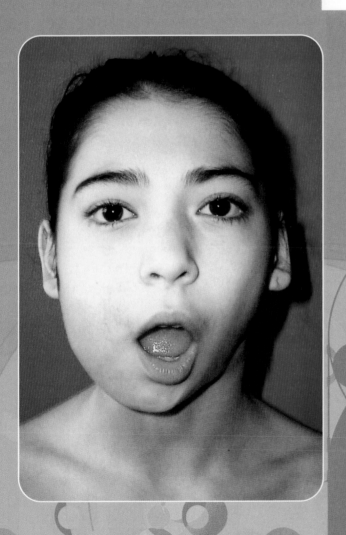

LEFT AND BELOW: *Anna Djantemiroma, before and after receiving the new polymer implant that has helped children with facial deformities or injuries to smile again.*

Hunting the Heart Attack Gene

Every human cell holds, within its nucleus, over 50,000 different genes. All of a person's genes come from his or her parents. Half come from the mother and half from the father via the egg and sperm cells. Each parent provides a different mix of his or her genes to each child, which accounts for the different appearance, personality and health among most brothers and sisters. Genes contain information for all aspects of bodily growth and development – influencing and directing, through the sum of their effects, the functioning of every organ and system in the body. Our genes determine our lives – from the colour of our hair, skin and eyes, to, in some cases, how long we will actually live.

Heart attacks are one of the biggest killers in modern society. Cardiac arrest can occur when the interior walls of the arteries carrying blood supply to the heart become clogged up, thereby impeding the flow of blood to the vital organ. The condition where the arteries narrow and harden is called atherosclerosis. It is atherosclerosis of the coronary arteries that causes most heart attacks. Although factors such as environment and lifestyle are heavy contributors toward heart disease, various genes also play a part. The number of genes associated with heart attacks could be more than 30, but in 2003 doctors in the United States discovered one that was prevalent in an extended family with a long history of heart disease. The Cleveland Clinic announced that it had identified the first gene confirmed as a cause of coronary disease in humans. It was promptly labelled "the heart attack gene".

The family in question were the Steffensens from Iowa. Their legacy of heart attacks began with the patriarch Arthur who died suddenly at 45 while driving. Years later, son Don was out duck hunting when his heart stopped Don survived, and initially accepted the view that the family history of heart disease was probably due to diet and exercise. He was concerned though that there might be an underlying reason why so many of his relatives had bad hearts.

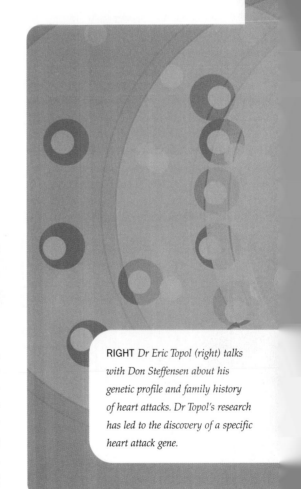

RIGHT *Dr Eric Topol (right) talks with Don Steffensen about his genetic profile and family history of heart attacks. Dr Topol's research has led to the discovery of a specific heart attack gene.*

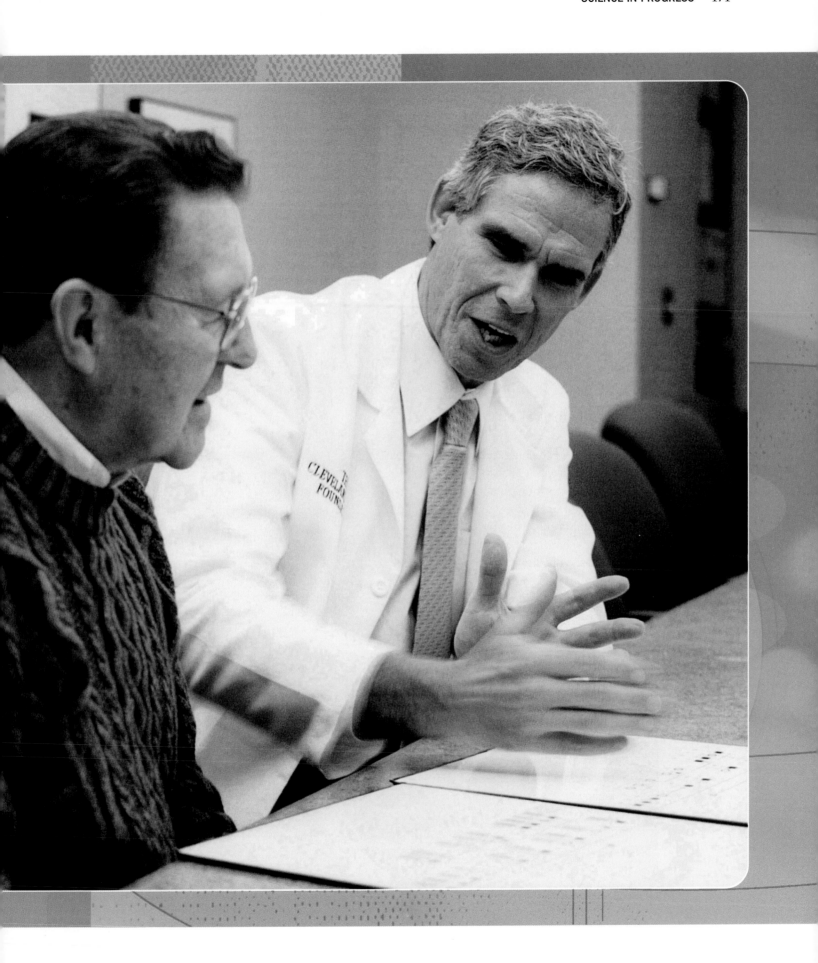

BELOW *Brothers Marvin (left) and Don Steffensen of Johnston, Iowa, pose with a family group photo taken in the early 1940s. Ten of the siblings, including Don, have had heart attacks.*

In 2002 he approached the Cleveland Clinic for help. It just happened that Dr Eric Topol, chairman of cardiology at the clinic, was, like others in his field, searching for a large family to study as part of the ongoing quest to discover a gene that causes heart attacks. While Don was at the clinic, he mentioned to one of the cardiologists that he had ten siblings, nine of whom had suffered heart attacks. When Dr Topol looked at their hospital records, he noticed that eight of the family had their heart attacks between the ages of 59 and 63, including four at 61 and two at 62. "It's very striking," he said. "Typically, heart attacks occur in the mid 50s to 60s, but to see the clustering almost at the same age among the family was another strong signal that this was a genetic story playing out."

For Dr Topol and his team of 50 researchers, the Steffensens were a gift from the gods. Not only was there rampant heart disease within the one family, but also there were plenty of heart attack survivors – including 76-year-old Elayne who had suffered three attacks – who were willing to give their blood so that their DNA could be examined. The Cleveland team spent over a year sifting through billions of bits of genetic information from each of the 21 participating Steffensens (only one declined to take part) and, by comparing the family members with heart disease to the lucky few who never had it, Dr Topol hoped to find the gene at the root of the disease. His eventual findings surprised him because it was a gene that nobody had ever connected to heart attacks or heart disease. The culprit was a mutation of the MEF2A gene.

The Cleveland scientists found that the MEF2A gene makes a protein that creates weak artery walls, allowing atheroslerotic plaques to form. Once blockages have occurred in the coronary arteries, a heart attack is a possible outcome. The defective gene was found in every member of the Steffensen family who had heart disease. "Even though this may turn out to be a rare condition," said Dr Topol, "it really sets the stage for unravelling the other key genes that cause heart attack and coronary disease." The discovery means that, perhaps by means of a simple blood test, doctors will be able to detect the gene in patients and advise them to be particularly careful about things like cholesterol and high blood pressure, both of which are known to encourage heart disease.

In fact, the "heart attack gene" is more common than Dr Topol initially suspected; the following year further studies at the clinic revealed that up to two per cent of all US heart attack and coronary artery disease patients may carry various mutations of the MEF2A gene.

Dr Qing Wang, director of the Cleveland Clinic's Center for Cardiovascular Genetics, said:

> **"Identifying new mutations in the MEF2A gene is a significant finding because it brings us one step closer to unlocking the genetics behind heart disease and heart attack. The finding that one per cent to two per cent of the heart disease population have mutations in MEF2A also is significant because it brings us closer to the development of a future genetic testing kit for these patients. Aggressive lifestyle modifications and preventive therapies will help patients prevent or delay the onset of heart attacks."**

As studies into coronary disease continued, genetic scientists in Stockholm announced that they had identified the first gene to link autoimmune and cardiovascular diseases. The autoimmune response – when the immune system mistakes the body's own tissues as foreign invaders and attacks them, causing inflammation – is at the root of diseases such as arthritis, diabetes and multiple sclerosis. Inflammation is a major contributory factor in atherosclerosis. Scientists at the Karolinska Institute found that a variant form of a gene called MHC2TA reduces the production of a number of proteins on which the immune system relies to fight diseases. A study of over 4,000 individuals – patients and healthy people – showed that heart attack victims were 39 per cent more likely to carry the mutant gene; arthritis sufferers were 29 per cent more likely to carry it; and people with MS were 14 per cent more likely to have it.

With around a million people a year dying in the United States alone from heart disease, and with strong evidence suggesting that this figure will rise considerably by 2020, any research that can prevent cardiac problems in future generations has to be welcomed.

The Pill That Could Add 30 Years to Your Life

Ever since medicine was first practised, the goal has been to find some method of prolonging human life. Now, according to a Scottish scientist, a pill could be on the market within the next decade that could extend our lives by as much as 30 years.

Professor John Speakman of Aberdeen University's zoology department believes that the elixir of life is to be found in a hormone called thyroxine, which is produced naturally in a gland in the throat. A daily additional dose – probably in pill form – could help us live to around 105. He bases this optimism largely on tests carried out on mice. When mice were given thyroxine, they lived considerably longer than might otherwise have been expected.

Professor Speakman thinks that the key to prolonging life could be found in the body's metabolism and the efficiency with which it burns up energy. Biologists have always assumed that an animal's lifespan was limited by the amount of energy it burned up. Thus, while a mouse has a fast metabolism and only lives for a handful of years, an elephant, with its slow metabolic rate, lives for 70 years. However, Professor Speakman discovered that by increasing the metabolic rate of a mouse, through a dose of thyroxine, its lifespan was actually prolonged, thereby blowing a hole in the old adage of "live fast, die young". In tests, the mice that were given thyroxine lived for around 25 per cent longer than those that weren't. Professor Speakman says that this would translate in a human to a difference of around 30 years.

He claims the reason for the unexpected findings is due to a quirk in the normal metabolic cycle that occurs every time we breathe, resulting in the production of highly reactive and unstable molecules of oxygen, known as free radicals. These toxic by-products stalk through cells causing damage to DNA and other essential building blocks that are involved in ensuring the body runs smoothly. The more errors that occur within cells, the more they begin to malfunction, resulting in body tissues ageing. But

Professor Speakman suggests that animals with faster metabolic rates actually produce fewer of these harmful free radicals. He says that by creating drugs to replicate the effect – extra thyroxine is already given to people who do not produce enough of the hormone naturally, so that they have a healthy metabolic rate – it will take longer for the accumulative damage caused by free radicals to occur. So, by altering this biological process, he believes he can postpone the process of ageing.

He told *Scotland on Sunday*:

"We found that making the metabolism in cells less efficient meant more oxygen was consumed and fewer free radicals were produced. In effect, a pharmaceutical target would pump up the metabolic rate to reduce the free radical production. Because free radicals also play a role in causing cancer and other diseases, reducing the number produced can not only increase the length of time someone lives, but increase their healthy lifespan. This means they will not be spending any extra years they gain in a nursing home, but will have it added to their middle life. We could see people retiring at the age of 80, so they will be productive for much longer. The social and economic impact of that would be profound."

The key to the success of the proposed thyroxine pill is getting the right dose. For people with too much

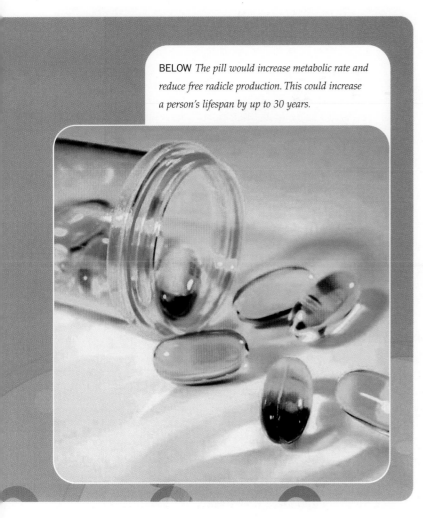

BELOW *The pill would increase metabolic rate and reduce free radicle production. This could increase a person's lifespan by up to 30 years.*

thyroxine in their bodies have to take medication to bring their metabolic level back to normal.

Funded with £450,000 from the Biotechnology and Biological Sciences Research Council, Professor Speakman plans to spend four years doing research with mice to determine the most effective dose of the drug. He says: "The grant will allow us to do experiments to see if we can put up the metabolic rate and get the correct thyroxine. We will put thyroxine in drinking water to get the metabolic rate up and then look at the difference in lifespan." He also intends to look into what stage in life people would need to start taking the drug.

A note of caution was sounded by Dr Pierre Bouloux, a leading specialist in human hormone disorders at the Royal Free Hospital in London. Doubting whether a mouse was the best model for a man in this instance, he suggested that because mice have a different metabolism to humans, the findings from one would not be relevant to the other. Only time will tell, and if Professor Speakman's predictions come true, time is something we might all have more of in the future.

Face Transplants: The Way Ahead?

Organ transplantation has long aroused curiosity and controversy, being praised by many as a lifesaving procedure but criticized by others for its perceived intrusion on personal identity. When, in 1998, surgeons in France conducted the first human hand transplant (pp. 96–102), it was the first time that a patient had taken on the risks of lifelong immune-suppressing drugs in order to keep a body part that was not needed for survival. A number of hand transplants have been performed since then, with varying degrees of success, along with other non-lifesaving procedures, including a larynx transplant that restored a patient's ability to speak and the first human tongue transplant. Full face transplants, which involve lifting an entire face from a dead donor (including nose cartilage, nerves and muscles) and transferring them to someone that has been hideously disfigured by burns or other injuries, present a whole new challenge, however.

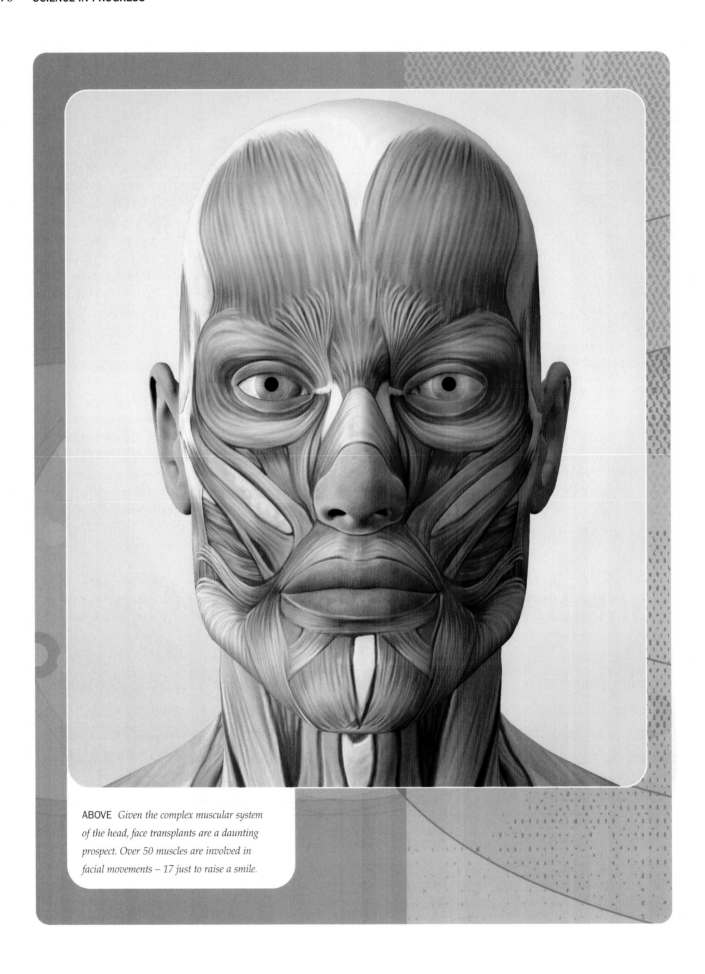

ABOVE *Given the complex muscular system of the head, face transplants are a daunting prospect. Over 50 muscles are involved in facial movements – 17 just to raise a smile.*

Faces have been replaced before. In 1994, a nine-year-old child in northern India lost her face and scalp in a threshing machine accident. Her parents raced to hospital with her face in a plastic bag and a surgeon managed to reconnect the arteries and replant the skin. People with terrible facial disfigurements, however, have generally had to rely on the grafting of small pieces of their own skin from another parts of their body. Some burns victims have endured over 50 skin graft operations to rebuild faces that they still consider to be hopelessly disfigured. Full face transplants would offer the hope of a better appearance and of better facial function.

The face is central to our understanding of our own identity. It is the most obvious feature that distinguishes us from those around us and it gives us our individuality. The face provides evidence of parentage, ancestry and racial identity. Most of our emotions, such as joy, anger or worry, are communicated through facial expressions, such as a smile, a snarl or a frown. To lose that messaging system can be socially catastrophic and therefore any transplant capable of restoring facial mobility would be welcomed. Faces have a number of types of skin with specialized characteristics – the inside of the eyelids and lips, for example, is not suitable for standard skin grafts since they do not allow for movement or sensitivity. For patients whose facial muscles are mostly intact, a transplant of an entire face, with its underlying fat and embedded blood vessels, could greatly enhance the ability to eat, drink and keep the eyes moist. And for patients with deeper disfigurement, there is the possibility that transplants could incorporate underlying facial muscles in order to restore that essential facial mobility.

Although surgeons in several countries are eager to carry out face transplants, the 24-hour operation is a daunting prospect. More than 50 muscles are involved in facial movements – it takes 17 muscles just to smile. For a full transplant surgeons need to save the donor's skin from hairline to jawline and ear to ear and also the nose, mouth and lips, eyebrows and eyelids, subcutaneous fat, some of the muscles, the nasal substructure and the nerves. Yet the greatest risk remains that of rejection. One of the most difficult tissues to transplant is skin because the body's first line of defence is especially sensitive to foreign tissue. This has restricted progress with transplants of external organs, such as hands. Because doctors do not know how violently the immune system will attack the face, there is the danger that the anti-rejection medicines will be inadequate. This problem is exacerbated by the fact that, in about 15 per cent of organ transplants, the patient becomes lazy about taking the drugs. If the new face is rejected, the resultant sloughing off would not only be disastrous emotionally, but it could also prove to be fatal.

There are also the psychological considerations of walking around wearing a dead person's face. As yet, doctors do not know how closely a transplanted face would resemble that of the donor, but if it should remind people of someone who is dead, that would pose an emotional problems. Peter Rowe, chairman of the ethical committee of the British Transplantation Society, said: "The main problem in these people is coming to terms with their new appearance. Since they would have to come to terms with a new appearance anyway, they may as well come to terms with their altered identity as with someone else's identity. Then there is the disfigurement of the potential donor. They were once living people and one should treat a corpse with respect."

This raises another potential pitfall. While there will probably be no shortage of willing recipients, donors could be harder to find. Families of potential donors could be reluctant to allow a transplant to proceed for fear of appearing disrespectful to the dearly departed. Others are concerned that if people come to believe that faces are "fair game" for procurement after death, fewer may volunteer to be organ donors. Therefore it has been agreed that only those who have given express permission should even be considered as face donors. Finding the right match for a transplant is equally problematic. Blood type, size and other biological factors all have to be taken into consideration, as a result of which any decision about who will donate and who will receive is likely to be a last-minute one. Families of the deceased may feel they need more time before making such a momentous decision.

Those in favour of pushing ahead with face transplants point out similar reservations before Dr Christiaan Barnard performed the first heart transplant in 1967. Whether or not full face transplants, that are seen as cosmetic not life saving, will become equally acceptable remains to be seen.

The Creation of Ozone in the Body

Scientists at the Scripps Research Institute in California, have recently discovered that the human body makes ozone. They say the body produces the gas in the course of a process that involves human immune cells called neutrophils and human immune proteins, or antibodies. It is thought that the presence of ozone in the human body may be linked to inflammation and that therefore this work could be of enormous significance with regard to the treatment of inflammatory diseases.

Ozone is a reactive form of oxygen that exists naturally as a trace gas in the atmosphere. It is probably best known for its crucial role in absorbing ultraviolet radiation in the stratosphere, where it is concentrated in the ozone layer and shields life on Earth from solar radiation. Ozone is also a familiar component of air in industrial and urban settings where it is a dangerous ingredient of smog. But ozone had never before been detected in biology.

Richard Lerner and Paul Wentworth of the Scripps Institute found that antibodies are able to produce ozone and other chemical oxidants when fed a rare,

BELOW *The ozone layer protects the Earth from solar radiation. But ozone is also a dangerous ingredient of the polluted smogs that form in the air over industrial regions.*

BELOW *The Scripps Research Institute, California, where scientists have recently discovered that the human body makes ozone. The presence of ozone may be linked to inflammation.*

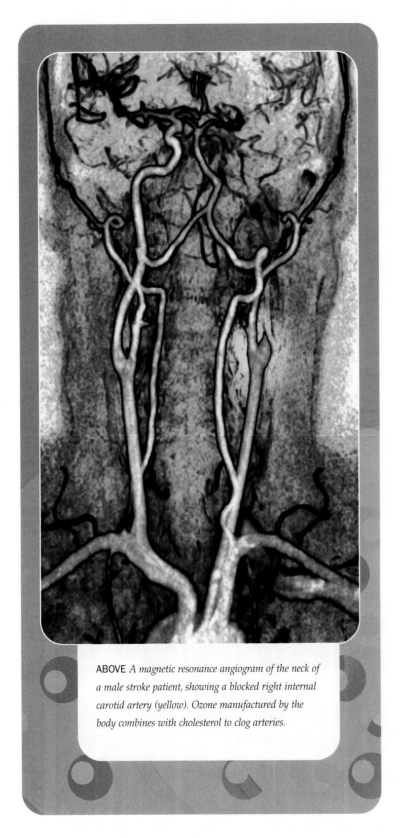

ABOVE *A magnetic resonance angiogram of the neck of a male stroke patient, showing a blocked right internal carotid artery (yellow). Ozone manufactured by the body combines with cholesterol to clog arteries.*

was completely unexpected as for a century immunologists had believed that antibodies – proteins secreted into the blood by the immune system – acted solely to recognise foreign pathogens and to attract lethal immune cells to the site of infection. However, the Scripps scientists were still unsure as to how the singlet oxygen came into contact with the antibodies.

Then, in 2003, Wentworth and Bernard Babior located what they believed to be the origin of the singlet oxygen – the oxidant-producing neutrophils. During an immune response, the neutrophils engulf and destroy bacteria and fungi by blasting them with these oxidants. The findings of the Scripps scientists suggest that the antibacterial effect of neutrophils is enhanced by antibodies. In addition to killing the bacteria themselves, the neutrophils feed singlet oxygen to the antibodies, which then convert it into ozone.

Further investigation revealed that ozone manufactured by the body joins forces with cholesterol to become a major cause in clogging arteries. Tiny puffs of ozone are produced in cholesterol-laden deposits, or "plaques", that plug arteries in the heart. The ozone then reacts chemically with the cholesterol, encouraging the plaques to grow and possibly to cause inflammation that further damages the arteries. It is hoped that this discovery will lead to a new method of detecting artery disease in the form of a blood test for the chemicals formed when cholesterol and ozone combine. Drugs that block ozone could also become a new weapon against blood vessel disease.

"We are investigating what impact this discovery will have on other inflammatory diseases," said Dr Wentworth, "such as rheumatoid arthritis, multiple sclerosis and Alzheimer's disease."

If ozone can be harnessed to play a key role in curing disease, it would join nitric oxide, another gas better known as an air pollutant, and the poisonous carbon monoxide in a growing list of gases that transmit messages and regulate functions in the human body. Dr Ferid Murad, who identified the role of nitric oxide, won the Nobel Prize for Medicine and his discovery led to the development of Viagra, the drug used for treating impotence.

Dr Wentworth predicted: "Ozone is going to be up there with nitric oxide and carbon monoxide as biological signalling agents."

reactive form of oxygen called singlet oxygen. They subsequently demonstrated that the oxidants produced by antibodies could destroy bacteria by poking holes in their cell walls. This development

A Vision of the Future

Corneal scarring and blindness affects about ten million people worldwide. When people are blinded following an accident or disease, there are many instances where only the cornea is damaged and the rest of the eye is functional. In some of these cases, vision can be restored via surgery, either by means of a simple procedure for glaucoma or by a more complicated corneal transplant.

Around 100,000 people receive transplants each year using human donor corneas, but there is a high risk of rejection and once a patient has rejected a first corneal graft, the chances of a successful retransplantation are very small. An added problem is that in some countries, such as India and China, people are reluctant to donate body parts. Even in the UK there is usually a wait of several months before a donor cornea becomes available. The alternative is to use an artificial cornea transplant, a procedure known as keratoprosthesis.

The process involves first assembling the artificial cornea around donor tissue then removing the

BELOW *A healthy human eyeball. Light passes through the cornea (white arch, far left), the eye's main light-focusing structure. Damage to the cornea can be repaired by means of a corneal transplant.*

patient's damaged cornea, repairing other structural defects that would impair vision, such as removing a cataract, and finally sewing the artificial cornea assembly in place. The idea is that the artificial cornea will succeed because it uses donor tissue as a "glue" sandwiched between the front and back plates of the device, thereby improving the artificial cornea's binding to the patient's eye. In traditional transplant surgery, the entire cornea is transplanted, and when complications develop, donor tissue can become cloudy, which prevents light from entering the eye properly – consequently vision deteriorates.

The concept of an artificial cornea is not a new one. It dates back to the early nineteenth century. Since the late 1940s a variety of polymers and techniques have been employed, but these have met with only limited success in human patients. So the quest has continued to find the ideal material.

In the late 1990s researchers from the Lions Eye Institute in Western Australia developed the world's first flexible artificial cornea. Made from a combination of soft plastics, it is considered an improvement on the rigid synthetic corneas that were previously available because it is flexible, like a real cornea, more robust and can be implanted in one piece rather than several sections. Dr Ming Wang, who has performed transplants with the new cornea at the Wang Vision Institute in Nashville, Tennessee, has claimed an 80 per cent success rate in early trials.

There have also been encouraging reports from the United States concerning a new artificial cornea, developed over a 15-year period by Harvard ophthalmologist Claes Dohlman. It is made of polymethyl methacrylate – the same material used in contact lenses. Nikon Sandulyak, a veterinary surgeon from the Ukraine, was blinded in 1966 after sustaining sever chemical burns to the cornea. He never gave up hope that one day he would see again, but his efforts to find suitable treatment proved fruitless until he moved to California in 2003 to be near his two daughters. In May 2004 he received the Dohlman keratoprosthesis during a two-hour operation in Sacramento. The day after surgery, the bandage over his right eye was removed and he saw his grown-up daughter Olena for the first time since she was five. Before surgery, he had only light perception, which for a seeing person is equivalent to noticing that a light has been turned on in a room when the eyes are closed. After surgery, his vision was restored to 20/30, enabling him to see all manner of things, from the numbers on a wall clock

to the details in faces. "For the past 38 years I have seen nothing," said Nikon. "Now I can see everything. I am surrounded by beautiful colours and people. Before my operation, I was only able to feel the faces of my grandchildren. Now I can see how beautiful they are."

Rosemary Collins from Chicago, Illinois, has also benefited from a Dohlman keratoprosthesis. She had lived with progressive blindness in both eyes from corneal disease and glaucoma. In fact, she had been legally blind in her left eye for 13 years, during which time several surgeries, including a corneal transplant, failed to correct her vision. Then, in the spring of 2004, she was fitted with an artificial cornea

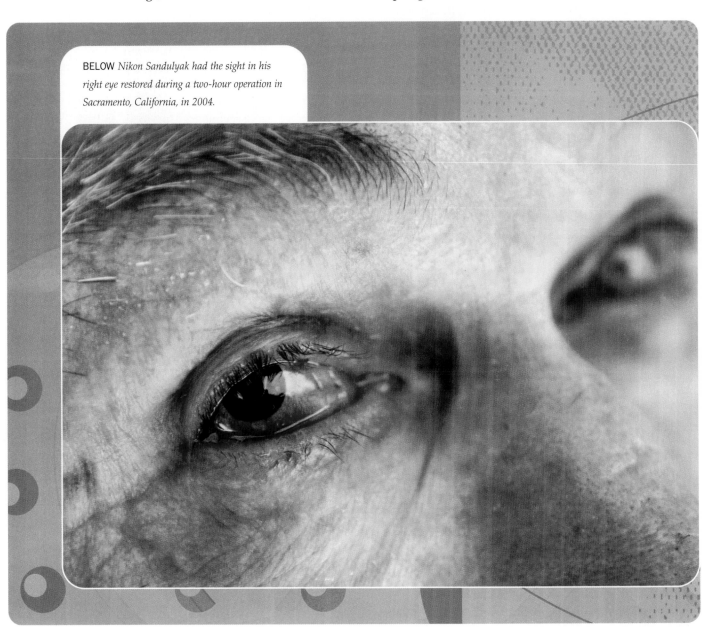

BELOW *Nikon Sandulyak had the sight in his right eye restored during a two-hour operation in Sacramento, California, in 2004.*

in her left eye. The first day after surgery she could see 20/60 when the bandage came off and could read 20/50. Two days later, her vision was 20/40 (legal to drive) and her reading was 20/20. Doctors were amazed by her progress.

She enthused:

" People used to bring me flowers so I could smell them, but now I can see them! It's the little things that get you in life. For instance, whenever I put toothpaste on the toothbrush, it would go all over. Or when I'd try to fill a glass of water or a cup of coffee, my vision was so bad I'd spill it. Now if I ask my family to do something for me, they say, 'You can do it yourself.' The only thing is, I'm afraid to close my eyes for fear I'll wake up and find that this is just a dream. "

For 200 years ophthalmologists have aspired to perfect an artificial cornea to reverse blindness. Now refinements in device design, combined with better drug therapy and long-term follow-up care, are improving the outlook and making artificial corneas a viable option for patients who have been unsuccessful with traditional corneal transplants.

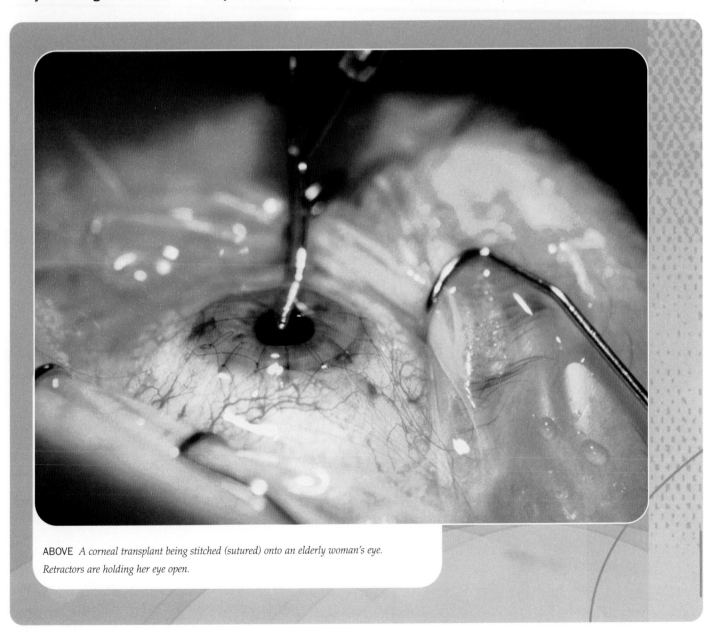

ABOVE *A corneal transplant being stitched (sutured) onto an elderly woman's eye. Retractors are holding her eye open.*

Lancaster University's Dr Nigel Fullwood, who is working on a British artificial cornea, insists that, despite the recent successes, there is still considerable scope for even more improvement. He is now hoping to make a cornea from a polymer with a high water content. He aims to modify the artificial cornea so that, during a transplat, it can be inserted in the same way as a conventional cornea, thus becoming fully integrated into the eye. His aim is to develop it by 2010 and predicts: "If we succeed, instead of waiting for a donor cornea, you will be able to get one off the shelf – in a similar way a plastic lens is used in a cataract operation."

Thanks to developments such as this, people with corneal scarring can finally see a brighter future ahead for themselves.

ABOVE *Nikon Sandulyak, with his wife Klavdiya (left) and youngest daughter Olena, talks about his gift of sight after an artificial corneal transplant.*

Scientists Predict Womb Transplant Baby

About 15 per cent of all couples are infertile. Most causes can be treated by *in vitro* fertilisation (IVF) and sperm injection. However, over 15,000 women in the UK who seek the help of fertility specialists each year are found to be incapable of getting pregnant because their womb has been destroyed, maybe as a result of a hysterectomy, cancer treatment or because they were born without one. Only around 200 opt for the emotionally draining path of IVF surrogacy, where their egg and their partner's sperm can be used, but another woman carries the baby. Surrogacy is also unacceptable in some cultures.

However, there could soon be a fresh opportunity for women to carry their own child – a womb transplant. Scientists have recently predicted that the first womb transplant baby may be just a few years away, but the whole concept has been dogged by controversy.

The world's first womb transplant actually took place as long ago as the year 2000 on a 26-year-old woman in Saudi Arabia. She had lost her own womb when a hysterectomy was carried out because of massive bleeding following a Caesarean section, and wanted another baby. The complex surgery, using the healthy uterus of a 46-year-old donor, was carried out at King Fahad Hospital in Jeddah. The transplant appeared to go well, after which the patient was given drugs to prevent her body rejecting the new womb. The drugs used were the same as those given to kidney transplant patients, many of whom have gone on to enjoy successful pregnancies. Stimulated by hormones, the womb's lining thickened to 18mm, which was more than enough to sustain a pregnancy. The new womb also restored the recipient's periods

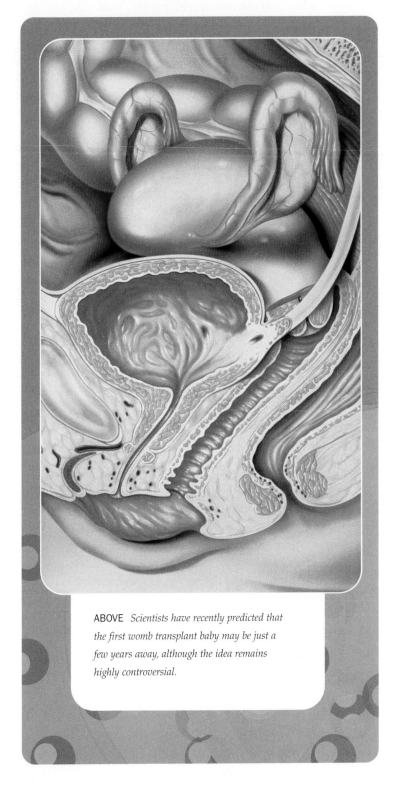

ABOVE *Scientists have recently predicted that the first womb transplant baby may be just a few years away, although the idea remains highly controversial.*

ABOVE *Fertility expert Lord Robert Winston is an outspoken opponent of womb transplants, arguing that, at present, they are dangerous both physically and ethically.*

for two menstrual cycles. Then, after 99 days, the surgeons had to remove the transplanted womb due to blood clotting, caused, it was thought, by the uterus moving within the pelvis.

Professor Wafa Fageeh, who led the transplant team, said that despite the complications the operation had been a "good start". Other gynaecologists agreed that the result was extremely promising. Richard Smith, consultant gynaecologist at the Chelsea and Westminster Hospital, told the *Guardian* newspaper: "They have achieved a lot. They have shown it is technically feasible to perform the operation in a woman". Addressing concerns that such a transplant would mean a lifetime on immunosuppressant drugs, he added: "Our view always was that the uterus would go in and the woman would have one or two babies and then the

uterus would come out – she would only be on immunosuppressive therapy for a few years".

Peter Bowen-Simpkins, from the Royal College of Obstetricians and Gynaecologists, said he believed the development would eventually lead to women without a uterus being able to give birth. "The womb survived for more than two menstrual cycles, so the first crucial hurdles have been passed," he said.

However, leading fertility expert Lord Winston strongly disputed that the Saudi Arabian transplant had been a success. Instead, he said, the blood clot proved that the entire procedure had been a total failure. "Putting a rotting piece of meat in the pelvis, which is what will happen with a womb transplant, will endanger the life of the recipient of the transplant and could cause a thrombosis. It would not be ethically justified in Britain or in the United States."

He added that it was a "terrible pity" that childless women's hopes had been raised by glowing reports of the operation. "Many women lose their uterus whilst still of childbearing age and there are other women who are born without a uterus. This simply will not help them."

One of Lord Winston's principal objections to the whole procedure was that, in 50 years of experiments, blood clotting had always prevented successful womb transplants, even in animals. Despite Lord Winston's misgivings, the step toward a human womb transplant baby moved a little closer in 2002 when Swedish scientists managed to produce a pregnancy in a womb transplanted into a mouse – the first time that a uterus from one animal had been transplanted into another and resulted in a successful pregnancy.   Professor Mats Brännström from Gothenburg University, who led the research, was confident that his success with mice paved the way for similar transplants on humans. He said: "Suitable donors could be either a sister, after she has had her own children, or a mother since the chance for a good immune and blood type match would be high. Then it would be possible to carry your own child in the same womb as you developed during your growth as a foetus." He went so far as to suggest that, technically at least, it might one day be possible to transplant a womb into a man, and use hormone injections to enable the pregnancy to succeed. For the time being at least, sceptics have had enough trouble getting their head around the idea of womb transplants in women without thinking about pregnant men!

The argument is that, unlike other organ transplants, a womb transplant would not be a matter of saving the recipient's life and that therefore it was not justifiable, particularly in view of the risks associated with taking immunosuppressant drugs. Nevertheless, the importance of a uterine transplant to the many thousands of women of childbearing age who have perfectly good ovaries, but no womb should not be underestimated. As US gynaecologist Louis G Keith wrote in the *International Journal of Gynaecology and Obstetrics*: "To some individuals, childbearing is the greatest event of a lifetime. To such persons, transplantation of organs of reproduction would not be considered frivolous or unnecessary, even though these organs do not sustain life."

Cloning Human Embryos

Advances in stem cell technology have been hailed as carrying potential cures for many crippling conditions, such as diabetes, spinal cord injuries, Parkinson's disease and motor neurone disease. Stem cells can be found in adults - we all have blood stem cells, which are located in the bone marrow and continuously replenish the body's red blood cells, white blood cells and platelets. However, those found in days-old embryos are far more prevalent and more easily manipulated into specialized cells, which could then be used to create cures or even replacement organs. As I have discussed elsewhere in this book, quite apart from the ethical consideration there are drawbacks to harvesting embryonic stem cells, but scientists still feel they hold the key to treating some of the world's most debilitating diseases.

One of the problems with embryonic stem cells is that which faces any transplant: tissue rejection. And this is where cloning comes in. Therapeutic cloning – the creation of human embryos solely for the production of stem cells, rather than the intention of creating a new human being – could be used to clone a patient's DNA, harvest stem cells and grow them into the type of tissue required. Scientists hope that this would eliminate the problems of tissue rejection caused when someone else's tissue is rejected in a transplant.

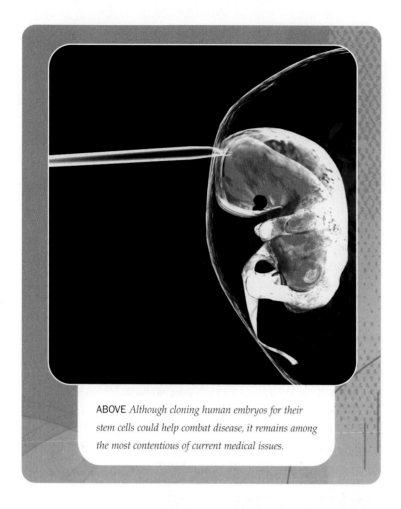

ABOVE *Although cloning human embryos for their stem cells could help combat disease, it remains among the most contentious of current medical issues.*

Cloning embryos for their stem cells would also provide a steady supply of cells for research purposes. But there remains widespread opposition from various bodies to any form of human embryo cloning. Anti-abortion groups and the Catholic Church have denounced embryonic stem cell research as "illegal, immoral and unnecessary". When Pope John Paul II visited the United States in 2001, he told President Bush that the work was as evil as infanticide. Washington has been pushing for a United Nations ban on all cloning, while other nations, led by Britain, are leading the call to allow cloning for medical experiments only.

It all started with Dolly the sheep. Back in 1997 Professor Ian Wilmut of the Roslin Institute in Edinburgh created the first successful mammal clone from an ordinary adult cell. The breakthrough sparked a host of imitators, with scientists at institutes across the world attempting to clone various species of animal while other scientists announced that they were working toward the first human clone. As the moral debate raged, the majority

of those involved in the research emphasized that the intention was not to create cloned human beings but to make lifesaving therapies. However, in 2001 Italian fertility doctor Severino Antinori and US researcher Panos Zavos caused a furore by announcing plans to clone humans. In the same year, the Human Fertilisation and Embryology Authority decided to make therapeutic cloning legal, with the stipulation that all embryos used for research must be destroyed when the work is finished. Indeed, the embryos are destroyed before they are 14 days old and are never allowed to develop beyond a cluster of cells the size of a pinhead. Nevertheless, many objectors still struggle to differentiate between therapeutic and reproductive cloning and express fears that therapeutic cloning could pave the way to allowing the creation of cloned babies. From the anti-abortion viewpoint, the argument is more straightforward: an embryo is a human life from the first moment of its existence and there is absolutely no justification for creating a human life for the sole purpose of experimenting on it.

In 2005 Professor Wilmut was granted a permit to clone human embryos in the hope of using stem cells extracted from them to treat Motor Neurone Disease (MND). Motor neurones carry signals from the brain to muscles around the body, but as the disease destroys them it causes muscle wastage. As the nerves degenerate, the disease invariably affects the muscles involved in breathing and swallowing, although the sufferer maintains intellect and awareness. In such cases MND is often fatal within two to four years of diagnosis.

Motor Neurone Disease was identified 130 years ago but scientists still do not know what causes it. A defect in a gene called SOD1 is known to account for two per cent of cases and a further eight per cent are known to be hereditary and therefore have some genetic basis. Professor Wilmut and his team are investigating a cloning technique called cell nuclear replacement, in which the nucleus of a human egg cell is removed and replaced with the nucleus from a human body cell, such as a skin cell. Since the replacement nucleus will come from patients with MND, the embryo will also have MND. The egg will then be stimulated to develop into an embryo. The embryo will be allowed to develop for around six days before being destroyed and during this time stem cells will be extracted and turned into the type of nerve cells affected by MND. As these cells grow, it will allow scientists to study, for the first time, the progress of the disease from the moment it takes hold until it finally destroys the cells. "It could be an extremely powerful tool for studying disease," said Professor Wilmut. "Our objective is to understand the disease. We hope one day it will lead to treatment."

The news was greeted with dismay and anger by opponents of cloning, who suggested that scientists should examine alternative methods of investigating MND, such as studying embryos that have been rejected from IVF use because they carry an inherited disease. But Professor Peter Braude, from the Centre for Preimplantation Genetic Diagnosis at King's College London, pointed out: "Unlike other genetic disorders like cystic fibrosis and Huntington's disease, for which stem cell lines have already been created from affected embryos following preimplantation genetic diagnosis, there is no other way of producing a motor neurone stem cell line other than using cloning techniques."

In May 2005, scientists from Newcastle, led by Professor Alison Murdoch, announced the first cloning of a human embryo in Britain. But the news was overshadowed by a report from South Korea, whose researchers are about two years ahead of the British teams, hailing a major breakthrough in tackling disease through human cloning. Scientists led by Professor Woo Suk Hwang of Seoul National University created stem cells that were an identical match to those of patients with diabetes and spinal injuries. Skin cells from 11 men and women were used to provide the donor DNA., this included nine people with spinal cord injuries, one who had juvenile diabetes, and one with a genetic immune condition. In every case, the genetic fingerprint of the stem cells that were created matched that of the donor. It is clearly a significant step forward, but at what price?

As the controversy over cloning human embryos escalates, Arthur Kaplan, medical ethicist and director of the University of Pennsylvania's Center for Bioethics, believes the discussion should be divided into two separate issues – therapeutic cloning and reproductive cloning. "I think the big question is: if you make this kind of thing in a dish, have you created a human life? Can you make something that people have strong moral views about in terms of destroying it, in order to benefit other people? That's going to be the key debate." And it looks like being one that will run for many years to come.

LEFT *Professor Woo Suk Hwang announces a major breakthrough in human cloning. He and his team created stem cells that were a perfect match to those of patients with diabetes and spinal injuries.*

INDEX

BIBLIOGRAPHY

Marvels and Mysteries of the Human Mind – ed. Alma E Guinness (*Reader's Digest*, 1997)
Medical Curiosities – Robert M. Youngson (Robinson, 1997)
Mysteries of the Mind – Reuben Stone (Blitz, 1993)
The Odd Body – Dr Stephen Juan (Collins, 1995)

The following newspapers, periodicals and websites also provided background information: *Daily Mail*; *Daily Record*; *Daily Telegraph*; news.bbc.co.uk; *New Scientist*; *New York Times*; *The Guardian*; *The Independent*; *The Mirror*; *The Observer*; *The Scotsman*; *The Times*; *Time* magazine; *Washington Post*.

PICTURE CREDITS